AF679464

Reinforced Concrete Design II
Homework and Exams

Peter I. Kattan

Petra Books
www.PetraBooks.com

Peter I. Kattan, PhD

Correspondence about this book may be sent to the author at one of the following two email addresses:

pkattan@petrabooks.com

info@petrabooks.com

Reinforced Concrete Design II Homework and Exams
ISBN: 979-8-3493-4244-8
All rights reserved. No part of this book may be copied or reproduced without written permission of the author or publisher.
© 2025 Peter I. Kattan

Preface

These are the homework sets and exam sets for the course Reinforced Concrete Design II that was taught by Dr. Peter I. Kattan. These homework and exam sets are a follow-up volume to the successful volume Reinforced Concrete Design II Lecture Notes that is available online and at fine bookstores.

This volume includes three homework sets and four exam sets. Solutions in detail are provided for some of these sets. In particular, detailed handwritten solutions are provided for four homework sets and for four exam sets. Note that these sets as well as the associated Lecture Notes are for the second course in reinforced concrete design. Another set of notes as well as homework and exams are available for the first course on reinforced concrete design.

The author makes these notes and sets available as free downloads as PDF files on the internet. However, the printed paperback copies are provided at a greatly reduced price at fine bookstores online and everywhere. The total size of this volume is about 56 pages. Not all the solutions are provided but enough is given to keep the student busy for hours.

May 2025 Peter I. Kattan

Reinforced Concrete Design II
Homework and Exams

Contents

Reinforced Concrete Design II
Homework and Exams

APPLIED SCIENCE UNIVERSITY
FACULTY OF ENGINEERING

Reinforced Concrete Design (2)

Homework 2

1. Using the ACI interaction curves, design a short tied column to carry loads of P_u = 1690 kN and M_u = 120 kN.m if f_c' = 21 MPa and f_y = 400 MPa. Bending is about the y-y axis.

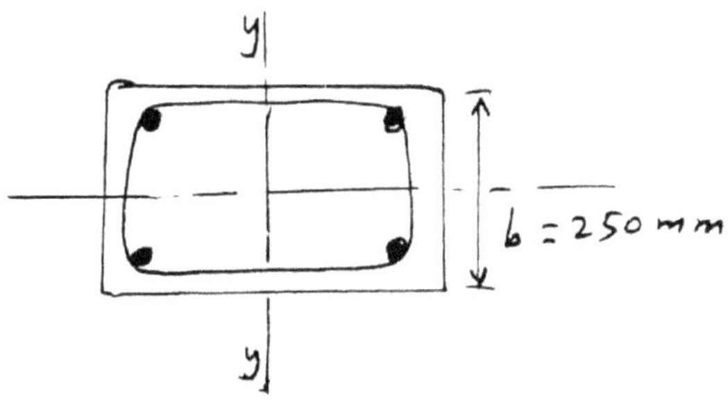

2. Using the ACI interaction curves, design a short circular spiral column to carry an axial force P_u = 2225 kN and a moment M_u = 128 kN.m. Take f_c' = 28 MPa and f_y = 400 MPa.

3. Using the trial method, determine the capacity of the short column with the cross-section shown in the figure for an axial load with an eccentricity of 125 mm. Use f_c' = 34 MPa and f_y = 400 MPa. Three φ25mm bars are on each side. Bending is about the y-y axis.

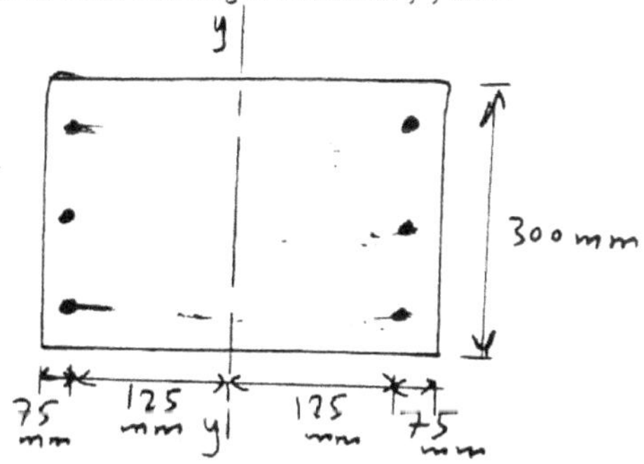

4. Determine whether the 300 x 350 mm tied column in a braced frame (see the figure) can be reinforced to support the indicated loads. Analysis shows bending about the y-y axis only, and 50 percent of the total factored axial load is due to dead load. Use k = 0.85, f_c' = 21 MPa and f_y = 400 MPa.

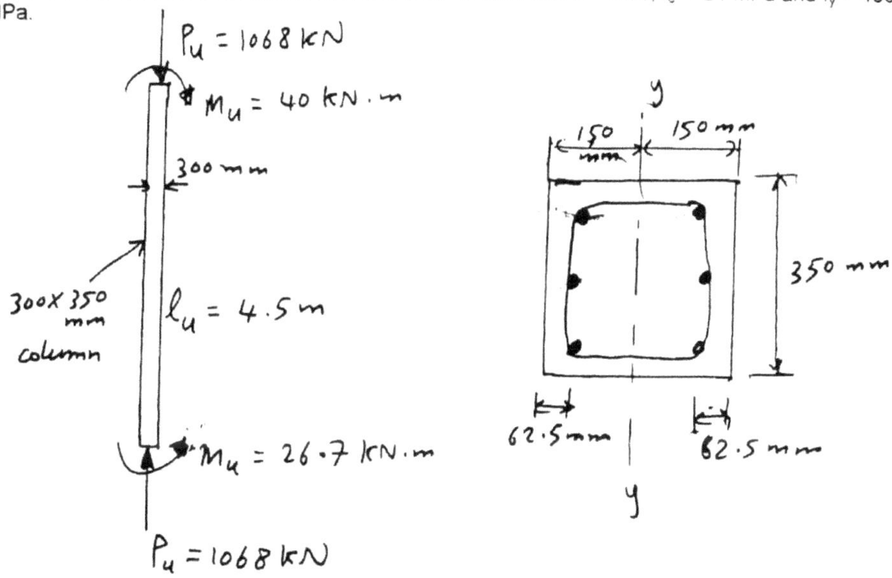

APPLIED SCIENCE UNIVERSITY
FACULTY OF ENGINEERING

Reinforced Concrete Design (2)

Homework 3

1. A wall footing (see the figure) is founded 1.8 m below grade on a soil with an allowable bearing capacity of 300 kN/m^2 at the base of the footing. If the footing supports a 250 mm thick concrete wall that carries a uniform load of D = 300 kN/m and L = 120 kN/m, design the footing. Take f_c' = 21 MPa, f_y = 400 MPa, and γ_s = 2162 kg/m^3.

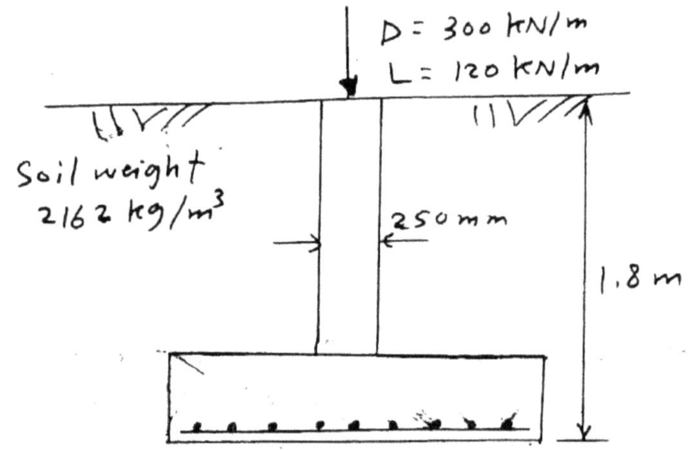

2. Design a square footing to support a 450 x 450 mm column (see the figure). The allowable soil pressure is 480 kN/m^2. Take f_c' = 21 MPa, f_y = 400 MPa, and γ_s = 2082 kg/m^3.

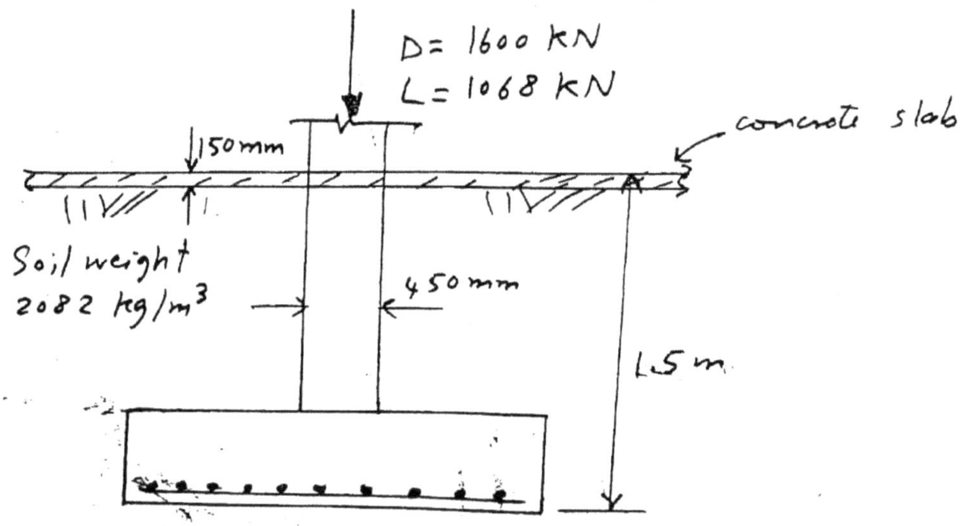

3. Design a rectangular footing to support a 400 x 400 mm column that carries service loads of D = 1335 kN and L = 1068 kN. The footing is 1.2 m below grade, and the width is not to exceed 2.1 m. The allowable soil pressure is 383 kN/m^2. Take f_c' = 24 MPa, f_y = 344 MPa, and γ_s = 2082 kg/m^3.

APPLIED SCIENCE UNIVERSITY
FACULTY OF ENGINEERING

Reinforced Concrete Design (2)

Homework 4

1. A torque T_u = 8000 kN.mm acts on the cross-section shown in the figure. Determine if torsion must be considered in sizing the reinforcement. Take f_c' = 30 MPa

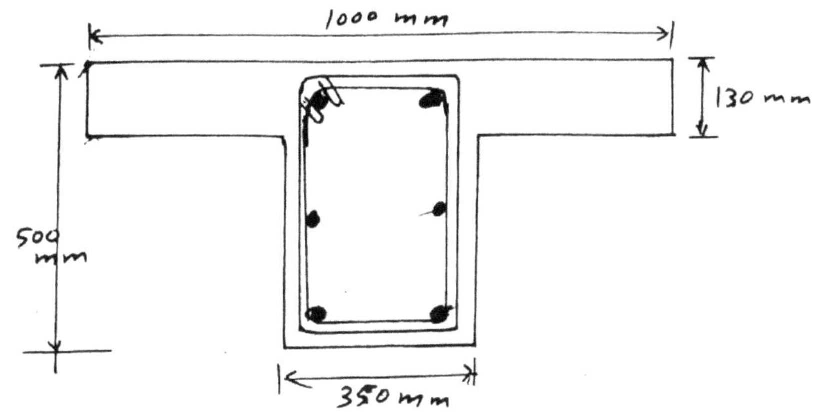

2. At a section of a beam the internal forces are V_u = 134 kN, M_u = 326 kN.m, and T_u = 136 kN.m. Will a rectangular cross-section 350 mm wide by 550 mm deep be adequate? If not, what width is required? Assume that the depth is to be held at 550 mm. Select the required bending and shear reinforcement and provide a design sketch. Take f_c' = 24 MPa and f_y = 400 MPa.

3. The internal forces acting on the cross-section of a beam are V_u = 160 kN, M_u = 270 kN.m (positive moment), and T_u = 120 kN.m. Assume clear cover of 40 mm for stirrups. Take f_c' = 21 MPa and f_y = 400 MPa.

   (a) Determine if the cross-section shown in Figure (a) can be reinforced to support the design loads.

   (b) If the design loads act on the cross-section shown in Figure (b), design the reinforcement and provide a design sketch.

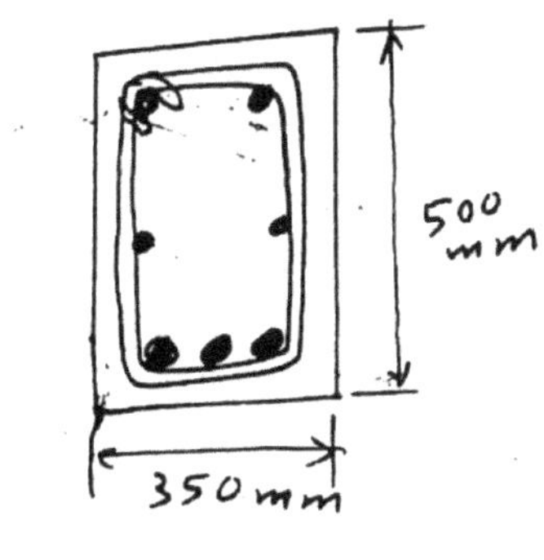

Figure (a)

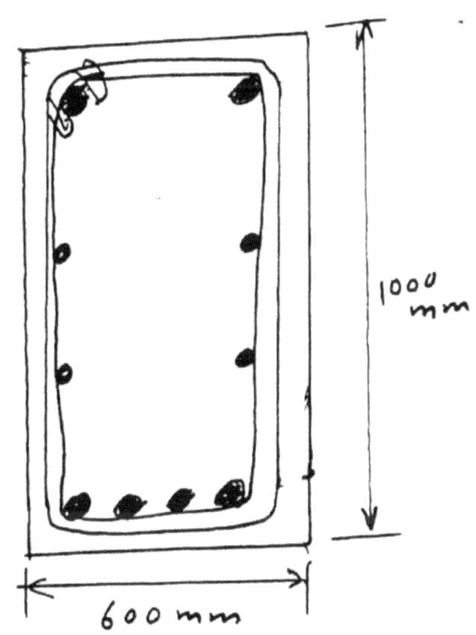

Figure (b)

Applied Science University
Faculty of Engineering

Dr. Peter I. Kattan
Time: 1 hour

Reinforced Concrete Design (2)
Exam 1

Answer all questions:

1- (50%) It is assumed that the column shown in the figure has a strain on its compression edge equal to -0.003 and has a tensile strain of +0.002 on its other edge. Determine the values of P_n and M_n which cause this strain distribution if f_y = 400 MPa and f_c' = 28 MPa.

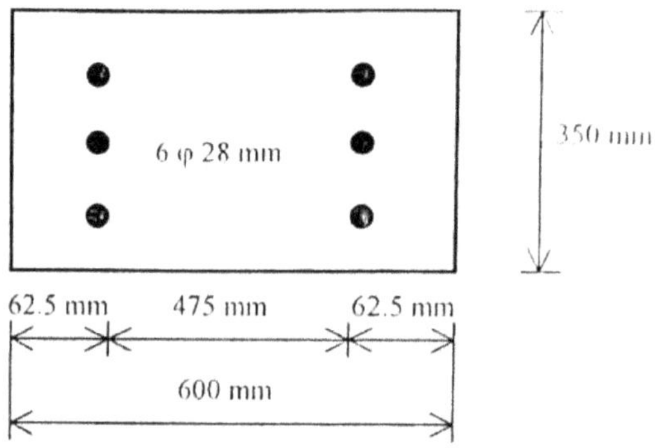

2- (50%) The tied column shown in the figure has been approximately sized to the dimensions 300 mm x 375 mm. It is to be used in a frame braced against sidesway. The column is bent in single curvature about its y-axis and has a clear height l_u = 4.8 m. If k = 0.83, f_y = 400 MPa (60 ksi), and f_c' = 28 MPa (4 ksi), determine the reinforcement required. Consider only bending in the plane of the frame. Note also that the dead axial load P_D is 134 kN.

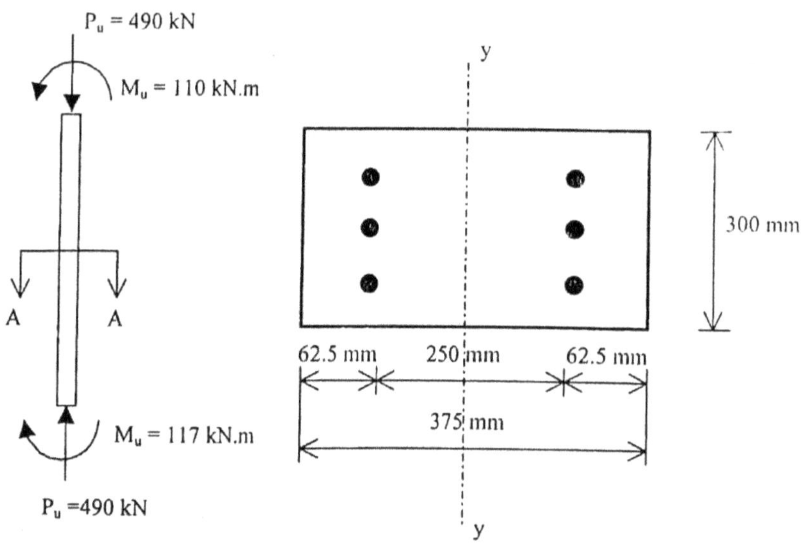

Section A-A

Applied Science University
Faculty of Engineering

Dr. Peter I. Kattan
Time: 1 hour

Reinforced Concrete Design (2)
Exam 1

1. (35%) (a) Using the alignment charts, calculate the effective length factor for column AB of the unbraced frame shown in the figure. Consider only bending in the plane of the frame.
   (b) Compute the slenderness ratio of column AB. Is it a short or slender column?

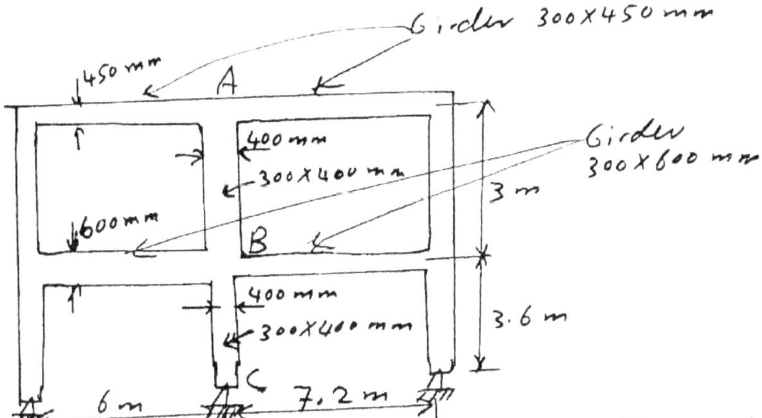

2. (30%) The 300 x 500 m short tied column shown in the figure is to be used to support the following: P_D = 445 KN, P_L = 490 kN, M_D = 81 KN.m, and M_L = 108 kN.m. If f_c' = 28 MPa (4 ksi) and f_y = 400 MPa (60 ksi), select the reinforcing bars to be used in two end faces by using the interaction curves.

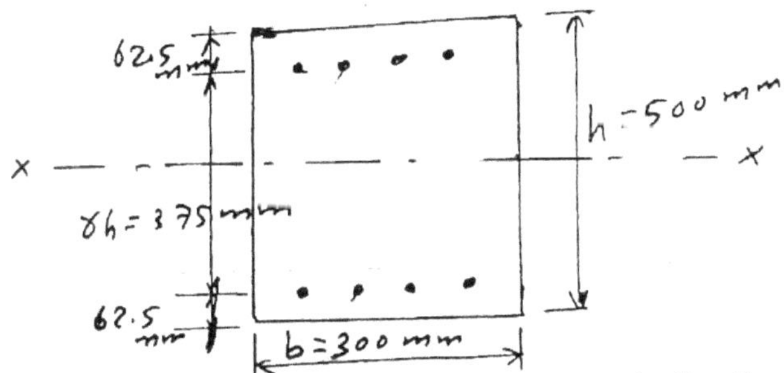

3. (35%) Determine the design capacity P_n of the short tied column shown in the figure which is subjected to biaxial bending. Use f_c' = 28 MPa (4 ksi), f_y = 400 MPa (60 ksi), e_x = 400 mm, and e_y = 200 mm. (Hint: Use the Bresler equation).

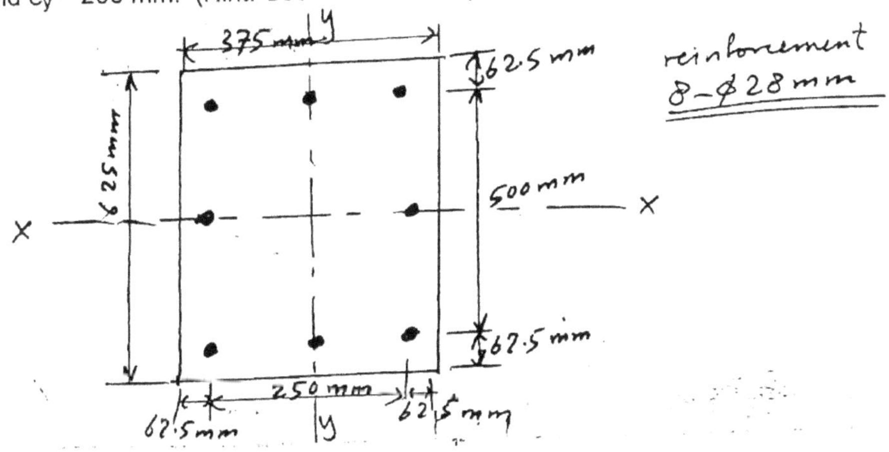

Applied Science University
Faculty of Engineering

Dr. Peter I. Kattan
Time: 1 hour

Reinforced Concrete Design (2)
Exam 2

1. (50%) The cross-section shown in the figure is loaded by V_u = 222 kN and a torque T_u = 40 kN.m. Take f_c' = 21 MPa and f_y = 400 MPa.

   (a) Check if torsion can be neglected.
   (b) What is the required spacing of φ12mm closed stirrups.
   (c) What area of longitudinal steel is required for torsion.

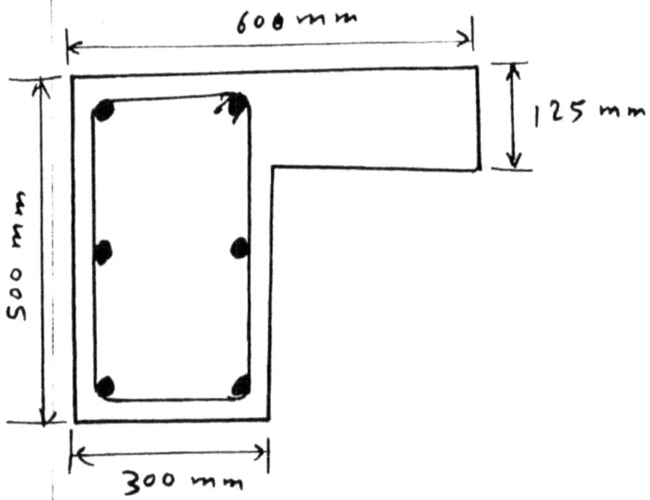

2. (50%) Consider a square column footing for a 400 mm square tied interior column that supports a dead load D = 890 kN and a live load L = 712 kN. The column is reinforced with 8-φ25mm bars, the base of the footing is D_f = 1.5 m below grade, the soil weight is γ_s = 1600 kg/m^3, f_y = 400 MPa, f_c' = 21 MPa, and the allowable soil pressure on the base of the footing is q_a = 240 kN/m^2.

   (a) Establish the dimensions of the footing.
   (b) Check punching shear (two-way shear) and beam shear (one-way shear).
   (c) Design the dowels for the transfer of load from column to footing.

Applied Science University
Faculty of Engineering

Dr. Peter I. Kattan
Time: 1 hour

Reinforced Concrete Design (2)
Exam 2

Answer all questions:

1- (100%) Consider a <u>rectangular</u> footing for a 450 mm square edge column with a dead load of 820 kN and a live load of 670 kN. Make the length of the long side equal to twice the width of the short side. Take f_y = 400 MPa, f_c' = 28 MPa, γ_s = 2000 kg/m^3, and the allowable soil pressure on the base of the footing is q_a = 200 kN/m^2. Assume the base of the footing is 1.5 m below grade.

(a) Establish the dimensions of the footing.
(b) Check punching shear (two-way shear) and beam shear (one-way shear).
(c) Determine the flexural reinforcement in the long direction.
(d) Determine the flexural reinforcement in the short direction.
(e) Provide a design sketch.

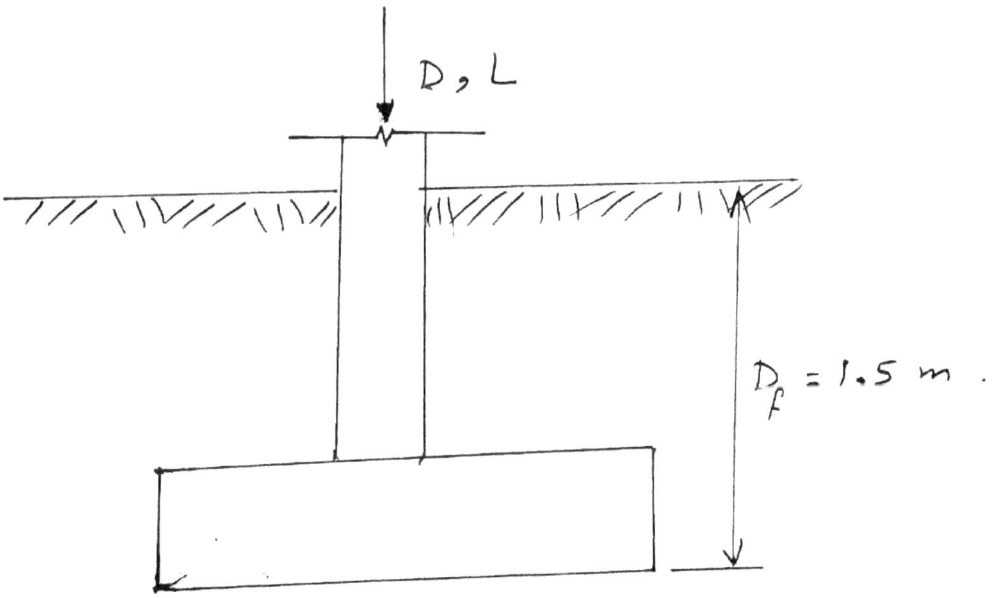

Reinforced Concrete Design (2)

Homework -1-

7.2 Classify column AB as long or short. Unbraced in plane of frame. All members 10" deep.

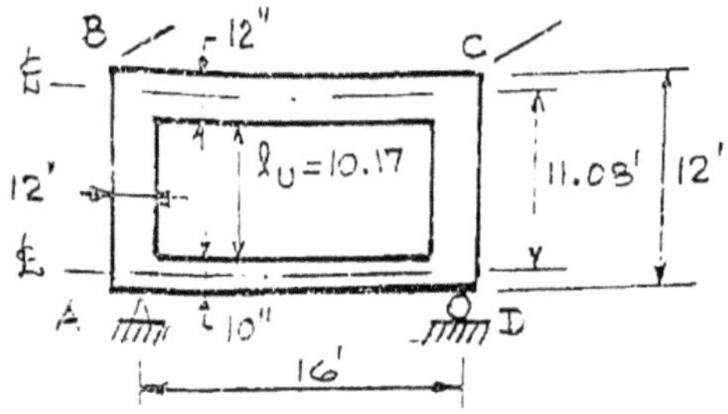

$r_{AB} = 0.3h = 0.3(12) = 3.6$

Adjust gross I's by values in Table 7.1

$I_{AB} = (0.7)\ \frac{10}{12}(12)^3 = 1008\ in^4$

$I_{BC} = (0.35)\ \frac{10}{12}(12)^3 = 504\ in^4;\quad I_{AD} = (0.35)\ \frac{10\ (10)^3}{12} = 292$

Joint B

$\psi_B = \frac{\sum \frac{E_c I_c}{L_c}}{\sum \frac{E_G I_G}{L_G}} = \frac{\frac{1008}{11.08}}{\frac{504}{16}} = 2.9$

READ IN Fig. 7.13b, k = 2

$\frac{k\ell_u}{r} = \frac{2(10.17 \times 12)}{3.6} = 67.8$

$\psi_A = \frac{\sum I_c/L_c}{\sum I_G/L_G} = \frac{\frac{1008}{11.08}}{\frac{292}{16}} = 5.0$

SINCE $\frac{k\ell_u}{r} > 22$ (EQ. 7.14)
AB is a "long" column

7.3 Classify columns AB and CD. all members 12" deep.
All beams $f_c' = 3ksi$, All columns $f_c' = 6\ ksi$

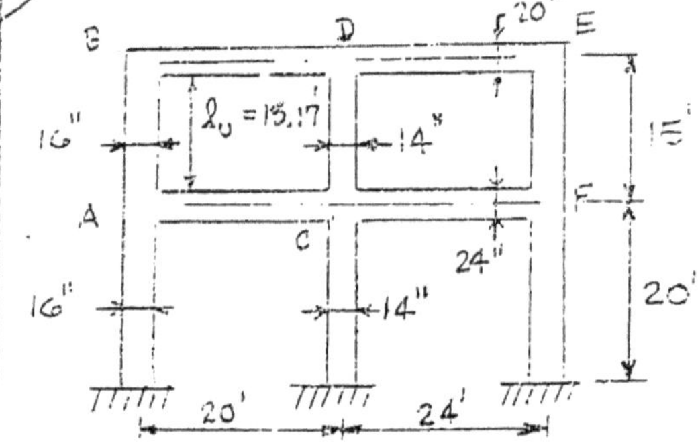

$E = 57{,}000\sqrt{f_c'}$

$E_c = 57{,}000\sqrt{\frac{6000}{1000}} = 4{,}415\ ksi$

$E_G = 57{,}000\sqrt{\frac{3000}{1000}} = 3{,}120\ ksi$

Adjust gross I's by values in Table 7.1

$I_{AB} = (0.7)\ \frac{12\ (16)^3}{12} = \underline{\underline{2867\ in^4}},\quad I_{CD} = (0.7)\ \frac{12\ (14)^3}{12} = \underline{\underline{1921\ in^4}}$

$I_{BD} = (0.35)\ \frac{12(20)^3}{12} = \underline{\underline{2800\ in^4}},\quad I_{AC} = (0.35)\ \frac{12(24)^3}{12} = \underline{\underline{4838.4\ in^4}}$

7.3 continued

(a) Braced frame   classify column AB

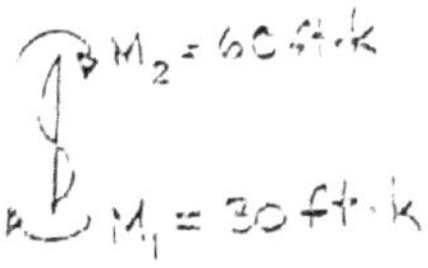

DOUBLE CURVATURE BENDING

$$\psi_B = \frac{\sum \frac{E_c I_c}{L_c}}{\sum \frac{E_G I_G}{L_G}} = \frac{\frac{4{,}415(2867)}{15}}{\frac{3{,}120(2800)}{20}} = 1.93$$

$$\psi_A = \frac{\sum \frac{E_c I_c}{L_c}}{\sum \frac{E_G I_G}{L_G}} = \frac{4{,}415\left[\frac{2867}{15} + \frac{2867}{20}\right]}{3{,}120\left[\frac{4838.4}{20}\right]} = 1.36$$

Read in Fig. 7.13(a) $k = 0.85$; $\frac{k\ell_u}{r} = \frac{0.85(13.17 \times 12)}{0.3(16)} = 28$

$$34 - 12\frac{M_1}{M_2} = 34 - 12\left(-\frac{30}{60}\right) = 40$$

SINCE $k\ell_u/r < 40$ COL AB is "SHORT"

(b) unbraced frame   classify col. AB

Use same values of ψ_A and ψ_B, READ IN Fig 7.13b $k = 1.58$

$$k\ell_u/r = \frac{1.58(13.17 \times 12)}{0.3(16)} = 52 > 22 \therefore \text{Col. AB CLASSIFIED AS "LONG"}$$

(b) classify column CD

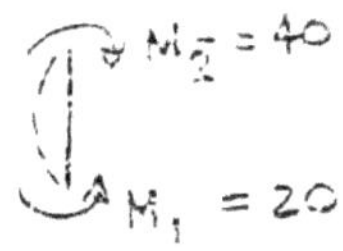

SINGLE CURVATURE BENDING

$$\psi_D = \frac{\sum \frac{E_c I_c}{L_c}}{\sum \frac{E_G I_G}{L_G}} = \frac{4{,}415\left[\frac{1921}{15}\right]}{3{,}120\left[\frac{2800}{20} + \frac{2800}{24}\right]} = 0.7$$

$$\psi_C = \frac{\sum \frac{E_c I_c}{L_c}}{\sum \frac{E_G I_G}{L_G}} = \frac{4{,}415\left[\frac{1921}{15} + \frac{1921}{20}\right]}{3{,}120\left[\frac{4838.4}{20} + \frac{4838.4}{24}\right]} = 0.77$$

If frame braced, $k = 0.73$; $\frac{k\ell_u}{r} = \frac{0.73(13.17 \times 12)}{0.3(14)} = 27.5$

$$34 - 12\frac{M_1}{M_2} = 34 - \frac{20}{40}(12) = 28\text{;} \quad \frac{k\ell_u}{r} < 28 \therefore \text{SHORT}$$

If frame unbraced

READ $k = 1.24$ $k\ell_u/r = \frac{1.24(13.17 \times 12)}{0.3(14)} = 46.7 > 22$

Column is LONG

7.5 Determine the $k\ell_u/r$ of column AB. $f'_c = 6$ ksi

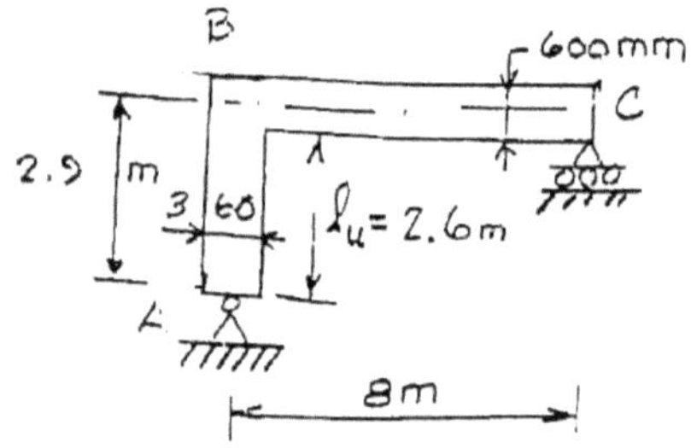

Use factors in Table 7.1 to modify moment of Inertia

Column AB

$$I_c = (0.7)\frac{300(360)^3}{12} = 8.16\times 10^8 \text{ mm}^4$$

Girder BC

$$I_G = (0.35)\frac{300(600)^3}{12} = 18.9\times 10^8 \text{ mm}^4$$

AT TOP OF COLUMN AB: $E_c = E_G$ ∴ CAN BE CANCELLED

$$\Psi_B = \frac{\sum \frac{E'_c I_c}{L_c}}{\sum \frac{E'_G I_G}{L_G}} = \frac{\frac{8.16\times 10^8}{2.9}}{\frac{18.9\times 10^8}{8}} = 1.19$$

AT BOT. OF COLUMN "AB"

$\Psi_A = \infty$ Pin end (see bottom of p 261 of text)

READ IN Fig 7.13b $k = 2.42$; $\ell_u = 2.9 - 0.3 = 2.6$ m

$h = 600$ mm $= 0.6$ m

$$\frac{k\ell_u}{r} = \frac{2.42(2.6)}{.3h} = 58.3 > 22$$

COLUMN AB CLASSIFIED AS "LONG"

(2) **7.6** Determine ϕP_n of cross section, short column

Use Eq 7.24a A_{ST}: 8-no.10 bars = 10.12 in^2 $f'_c = 6$ ksi

$$\phi P_n = (0.8)(0.7)[0.85f'_c(A_g - A_{ST}) + f_y A_{ST}]$$

$$\phi P_n = 0.56[.85(6)(12\times 18 - 10.12) + 60(10.12)] = 928 \text{ kips}$$

(3) **7.7** Locate the plastic centroid; $f'_c = 4$ ksi, $f_y = 60$ ksi

$A_s = 3.12$ in^2 $A_s = 1.2$ in^2

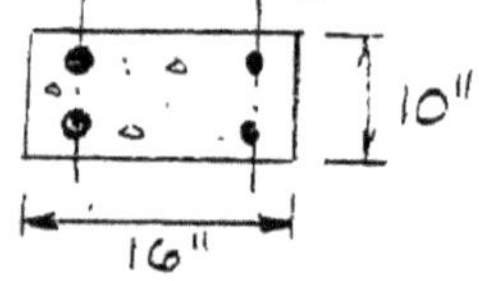

$F_c = 0.85f'_c A_G = 3.4(10\times 16) = 544^k$

$F_{S1} = 1.2(60 - 3.4) = 67.9^k$

$F_{S2} = 3.12(60 - 3.4) = 176.6^k$

COMPUTE P_n; $\Sigma F_y = 0$

$P_n = 544 + 67.9 + 176.6 = 788.5^k$

LOCATE P_n; $\Sigma M_o = 0$

$P_n \cdot \bar{x} = 67.9(13) + 544(8) + 176.6(3)$

$\bar{x} = 7.31$ in

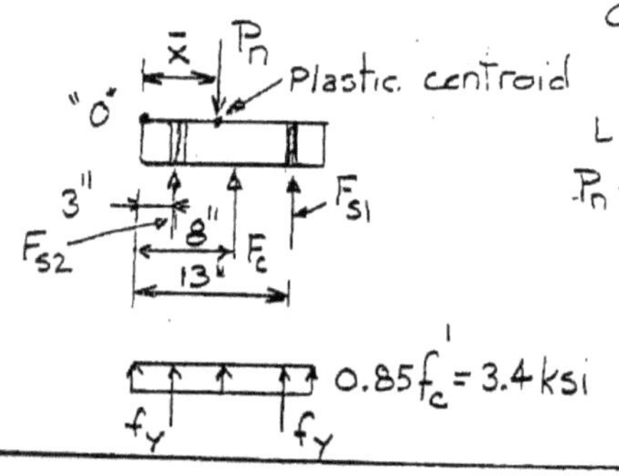

7.8

same figure as problem 7.7, but $f'_c = 6$ ksi, $f_y = 60$ ksi

$F_c = 0.85 f'_c A_G = 0.85(6)(10\times16) = 816^k$

$F_{S1} = A_{S1}(f_y - 0.85f'_c) = 1.2(60 - 0.85(6)) = 65.9^k$

$F_{S2} = A_{S2}(f_y - 0.85f'_c) = 3.12[60 - 0.85(6)] = 171.3^k$

$P_n = F_c + F_{S1} + F_{S2} = 816 + 65.9 + 171.3 = 1053.2^k$

LOCATE POSITION OF P_n; ΣM about left edge $= 0$

$\bar{x} P_n = 8F_c + 3F_{S2} + 13 F_{S1}$

$\bar{x} = 7.5$ in

7.9 (4)

CONSTRUCT INTERACTION DIAGRAM FOR 12"×14" TIED COLUMN. 4-#9, $f'_c = 4$ ksi, $f_y = 60$ ksi

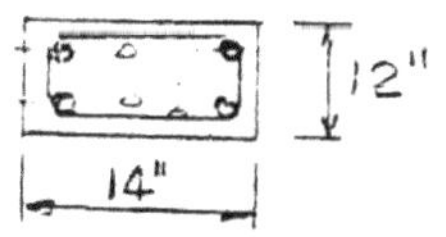

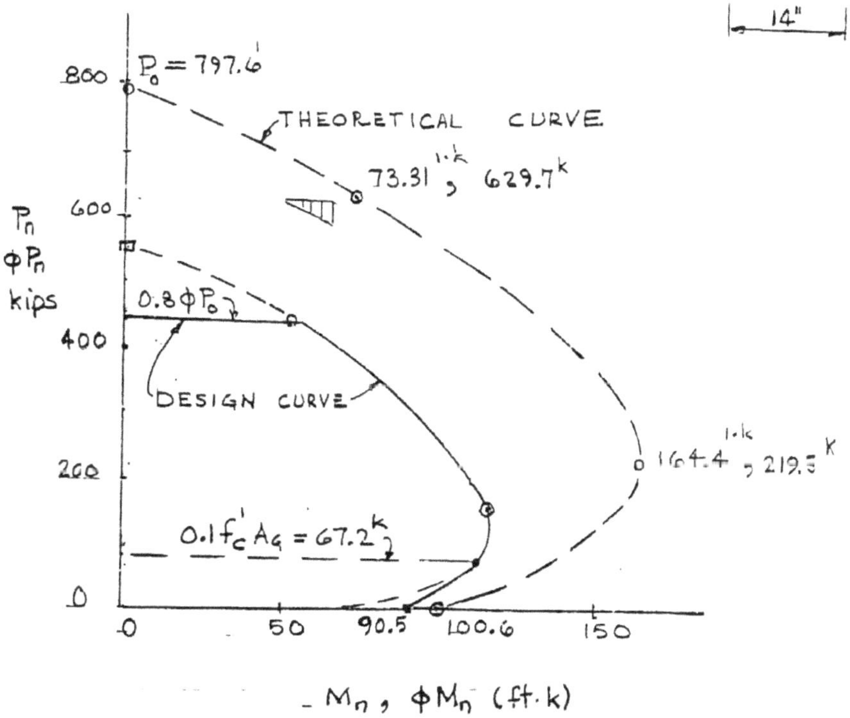

Reinforced Concrete Design (2)

Homework -2-

(1) **7.16** DESIGN A SHORT RECT. COLUMN TO CARRY $P_u = 380^k$, $M_u = 90$ ft·k. $f'_c = 3$ ksi, $f_y = 60$ kips/in². Use a trial method.

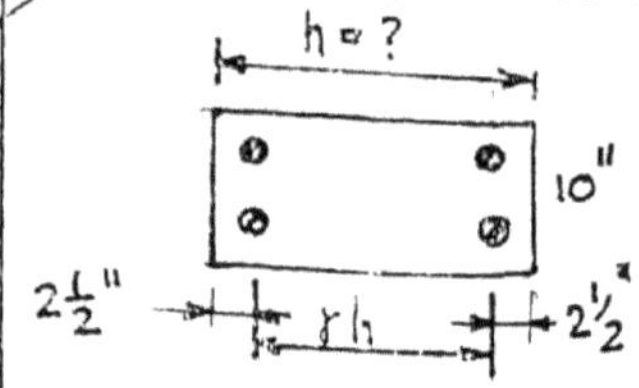

Estimate gross area, A_G. Assume average stress on cross-section $f_c = 2$ ksi

$$A_G = \frac{P_u}{f_c} = \frac{380}{2} = 190\ in^2 \quad \text{try } 10\times20;\ h = 20"$$

$$\gamma h = 20 - 2.5(2) = 15";\quad \gamma = \frac{15}{h=20} = 0.75$$

Use Fig. 7.39

Compute variables to enter design aid

$$e = \frac{M_u}{P_u} = \frac{90(12)}{380} = 2.84\ in \qquad \frac{\phi P_n}{A_G} = \frac{P_u}{A_G} = \frac{380^K}{200} = 1.9\ ksi$$

$$\frac{\phi P_n}{A_G}\times\frac{e}{h} = 1.9\ \frac{2.84}{20} = 0.27$$

Read in Fig. 7.39 $\rho_g = 0.02$; $A_{st} = \rho_g A_g = .02(200) = 4\ in^2$
10×20 — Use - 4 - no. 9 bars

Another design

NOTE: If $h = 18"$, $\rho_g = 0.031$, $A_{ST} = 0.031(180) = 5.58\ in^2$
10×18 — Use 4 - no. 11 bars

(2) **7.17** DESIGN A SHORT SPIRAL COLUMN TO CARRY $P_u = 500^K$, $M_u = 96'^{\cdot K}$; $f'_c = 4\ K/in^2$; $f_y = 60\ K/in^2$

ESTIMATE A_G, GUESS $f_c = 2.4\ K/in^2$

$$A_G = \frac{P_u}{f_c} = \frac{500^K}{2.4} = 208.3\ in^2,\quad \text{TRY 16"DIA, } A_G = 201\ in^2$$

COMPUTE VARIABLES REQUIRED FOR DESIGN AIDS

$$e = \frac{M_u}{P_u} = \frac{96\times12}{500} = 2.3\ in \qquad \frac{\phi P_n}{A_G} = \frac{P_u}{A_G} = \frac{500}{201} = 2.5\ K/in^2$$

$$\frac{\phi P_n}{A_G}\times\frac{e}{h} = 2.5\ \frac{2.3}{16} = .36 \qquad \gamma h = 11",\ \text{FOR } h = D = 16",\ \gamma = .69$$

INTERPOLATING BETWEEN FIGS 7.38 and 7.39 $\rho = .033$

$A_{ST} = \rho A_G = .033\times201 = 6.7\ in^2$, USE 7 - #9 ($A_{ST} = 7\ in^2$)

7.18

DESIGN A SHORT SPIRAL COLUMN TO CARRY $P_U = 950^K$, $M_U = 45$ ft·K, $f'_c = 4$ KSI, $f_y = 60$ KSI

SEVERAL TRIALS INDICATE $D = 18$ in REQUIRED

$A_G = 254.47$ in^2. $\gamma h = 18-5 = 13$, FOR $h = D = 18$ in

$\gamma = .722$, FOR $\frac{\phi P_n}{A_G} = 3.74$ AND $\frac{\phi P_n}{A_G} \times \frac{e}{h} = .118$

READ IN FIG 7.39 $\rho = .043$, $A_{ST} = \rho A_G = 10.9$ in^2

USE 7-#11, $A_{ST} = 10.94$ in^2

7.19

USE THE TRIAL METHOD OF SECTION 7.9 TO ESTABLISH THE CAPACITY OF THE CROSS SEC FOR AN AXIAL LOAD THAT ACTS WITH AN ECCENTRICITY = 5"

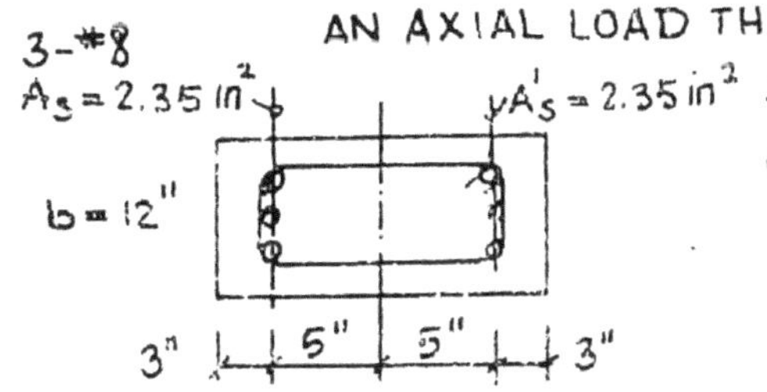

$f'_c = 5$ K/in^2; $f_y = 60$ K/in^2

$\epsilon_y = \frac{60}{29 \times 10^3} = .00207$

$\beta_1 = 0.85 - (f'_c - 4) = 0.8$

ONLY COMPUTATIONS FOR FINAL TRIAL SHOWN

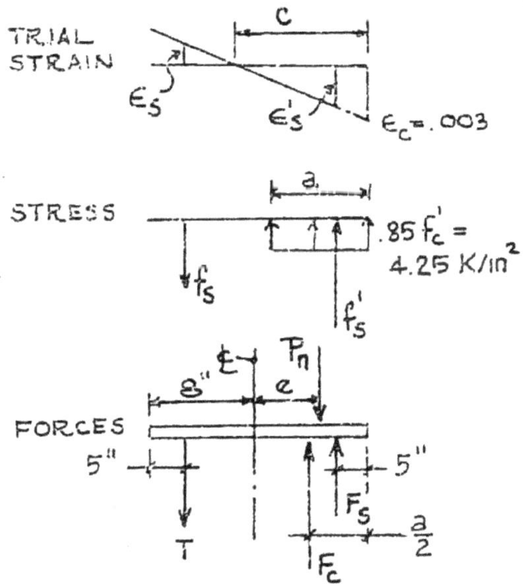

TRY $c = 10.3$ in, $a = \beta_1 c = .8(10.3) = 8.24"$

$\epsilon'_s = \frac{(c-3)\,0.003}{c} = .00212 > \epsilon_y$

$\therefore f'_s = f_y = 60$ K/in^2

$F'_s = A'_s(f_y - 0.85 f'_c) = 2.35(60 - 4.25)$

$= 131$ KIPS

$\epsilon_s = \frac{(2.7)\,0.003}{c} = .000786$

$f_s = \epsilon_s E_s = .000786 \times 29 \times 10^3 = 22.8 \frac{K}{in^2}$

$T = A_s f_s = 2.35(22.8) = 53.6^K$

$F_c = ab\,.85 f'_c = 8.24(12)\,4.25$

$= 420.24^K$

$\Sigma F_y = 0$, TO COMPUTE P_n

$P_n = F'_s + F_c - T = 131 + 420.24 - 53.6 = 497.64$

$\Sigma M = 0$ ABOUT ℄ TO ESTABLISH e

$0 = P_n \times e - F'_s 5 - F_c(8 - \frac{a}{2}) - T5$

$0 = 497.64e - 131 \times 5 - 420.24(8 - \frac{8.24}{2}) - 53.6 \times 5$

$e = 5.13" \approx 5"$ ∴ CLOSE TO SPECIFIED e. SAY OK

$P_U = \phi P_n = .7(497.64) = 348.6^K$

(4) **7.23**

CAN CROSS SECTION BE REINFORCED TO CARRY THE APPLIED LOADS? $k = 0.85$

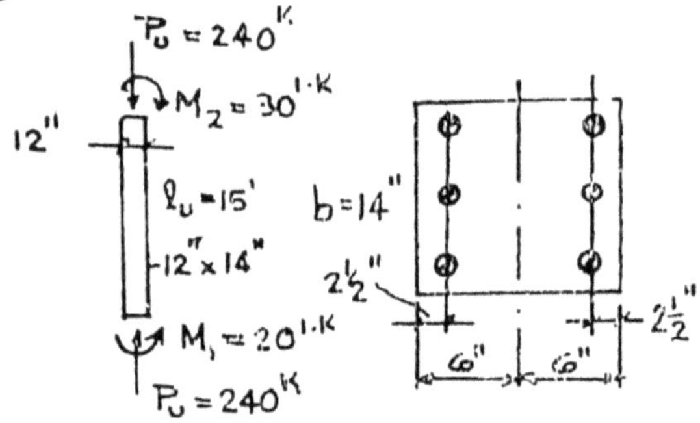

$\beta_d = \frac{M_D}{M_T} = .5$ $\qquad f'_c = 3{,}000 \text{ lb/in}^2$, $f_y = 60{,}000 \text{ lb/in}^2$

$\gamma h = 7"$, $\gamma = \frac{7}{12} = .58$ say $.6$

USE FIG. 7.38

DETERMINE IF COLUMN IS LONG OR SHORT

ACTUAL $\frac{K l_u}{r} = \frac{.85(15 \times 12)}{.3(12)} = 42.5$

$34 - 12\frac{M_1}{M_2} = 34 - 12\frac{20}{30} = 26$; $\quad 42.5 > 26 \therefore$ LONG

MAGNIFY M_2

MIN $M_2 = P_u(.6 + 0.03h)\text{in} = 240\left(\frac{.6 + 0.03[12]}{12}\right) = 19.2 \text{ ft}\cdot\text{K}$

USE $M_2 = 30 \text{ ft}\cdot\text{K} > 19.2 \text{ ft}\cdot\text{K}$

$M_c = \delta M_2$, $\quad C_m = .6 + .4\frac{M_1}{M_2} = .6 + .4\left[\frac{20}{30}\right] = .87 > .4$

$E_c = 57{,}000\sqrt{f'_c} = \frac{57{,}000\sqrt{3000}}{1000} = 3122 \text{ K/in}^2$

$I_G = \frac{bh^3}{12} = \frac{14(12)^3}{12} = 2016 \text{ in}^4$

$P_c = \frac{\pi^2 EI}{(K l_u)^2}$, $\quad EI = \frac{E_c I_G/2.5}{1 + \beta_d} = \frac{\frac{3122(2016)}{2.5}}{1 + .5} = 1{,}678{,}387 \text{ K}\cdot\text{in}^2$

$= \frac{\pi^2(1{,}678{,}387)}{\{.85(15 \times 12)\}^2} = 707.6^K$

Evaluate Eq. (7.32)

$$\delta_{ns} = \frac{C_m}{\left[1 - \frac{P_u}{0.75P_c}\right]} = \frac{.87}{\left[1 - \frac{240}{0.75(707.6)}\right]} = 1.59$$

$M_c = \delta_{ns} M_2 = 1.59(30) = 47.7 \text{ ft}\cdot\text{k}$

DETERMINE A_{ST} USING ACI INTERACTION CURVE: FIG 7.38

$\frac{\phi P_n}{A_G} = \frac{240^K}{12 \times 14} = 1.43 \text{ K/in}^2$ $\qquad e = \frac{M_c}{P_u} = \frac{47.7 \times 12}{240} = 2.39"$

$\frac{\phi P_n}{A_G} \times \frac{e}{h} = 1.43 \times \frac{2.39}{12} = 0.29 \text{ K/in}^2$ $\qquad$ READ $\rho = .015$

$A_{ST} = \rho bh = .015(12 \times 14) = 2.52 \text{ in}^2$; USE 4-#8

A_{ST} SUPPLIED $= 3.14 \text{ in}^2$

Reinforced Concrete Design (2)

Homework -3-

(1) 8.2 D = 20 K/FT, L = 8 K/FT

DESIGN THE ONE-WAY WALL FTG. $f'_c = 3{,}000\ lb/in^2$, $f_y = 60{,}000\ lb/in^2$, $\gamma_s = 135\ lb/FT^3$, $\gamma_c = 150\ lb/FT^3$

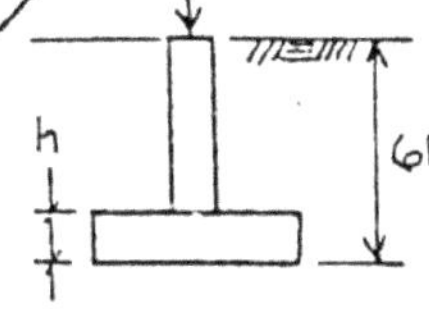

BASE ANALYSIS ON A ONE-FOOT SLICE OF WALL.

GROSS ALLOWABLE SOIL PRESSURE ON BASE OF FOOTING = 6 K/FT²

NET ALLOWABLE SOIL PRESSURE ASSUME h = 12 in

$q_N = 6 - (.15 + 5 \times .135) = \underline{5.18\ K/FT^2}$

WIDTH OF BASE TO LIMIT SOIL PRESSURE

$q_N = \frac{P}{A} = \frac{28^K}{b \times 1}$, $b = \frac{28^K}{5.18\ K/FT^2} = 5.4\ FT$ ROUND TO $\underline{5'-6''}$

q_u (FACTORED SOIL PRESSURE) $= \frac{P_u}{A}$

$= \frac{20 \times 1.4 + 8 \times 1.7}{5.5 \times 1} = \underline{7.56\ K/FT^2}$

CHECK SHEAR ON PLANE d INCHES OUT FROM FACE OF WALL. TRY h = 12 in

$d = h - (3\ IN\ COVER + \frac{d_b}{2})$ ASSUME $d_b = 1\ IN$

$d = 12 - 3.5 = 8.5\ IN$; THEN $a = 33 - 8.5 = 19.5''$

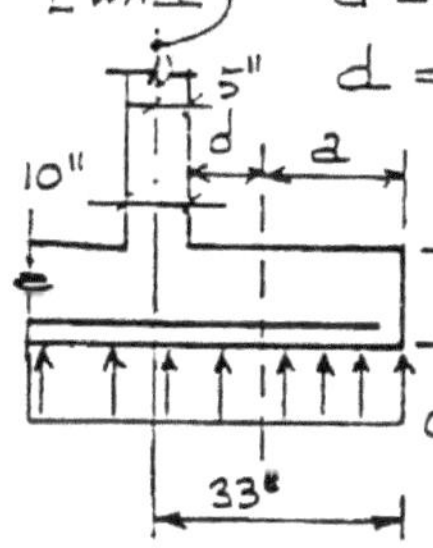

SHEAR ON CRITICAL PLANE

$V_u = \frac{19.5}{12} \times 1 \times 7.56 = \underline{12.29^K}$

SHEAR STRENGTH OF CONC.

$\phi V_c = .85(2\sqrt{f'_c}\ b_w d)$

$= .85(2\sqrt{3{,}000}\ \frac{12 \times 8.5}{1{,}000}$

$= \underline{9.5^K}$

SINCE $V_u > \phi V_c$, SECTION OVERSTRESSED, INCREASE h. TO 14"

NOW $d = 14 - 3.5 = 10.5\ IN$. REPEAT OF COMPUTATIONS GIVES $V_u = \frac{17.5}{12}(7.56) = 11.03^K < \phi V_c = 11.73^K$ ∴ OK

A_s REQ'D FOR MOMENT ; CRITICAL SECTION AT FACE OF WALL (USE TRIAL METHOD)

$M_u = \frac{28 \times 7.56}{12} \times \frac{14}{12} = 20.58\ FT\cdot K/FT$

$M_u = \phi T(d - \frac{a}{2})$ GUESS $a = 1\ IN$

$20.58 \times 12 = .9 A_s 60(10.5 - 0.5)$ ∴ $A_s = \underline{0.457\ IN^2/FT}$ GOVERNS.

CHECK a, EQUATE T = C

$A_s f_y = .85 f'_c\ a b$

$.457(60) = .85(3)\ a\ 12$

$a = .9\ in < 1\ in$ ∴ O.K.

8.2 cont.

Also check minimum steel

$A_{s,min} = 0.0018 A_g = 0.0018 \times 12 \times 14 = 0.30 \text{ in}^2/\text{ft}$

Moment governs ∴ Use $A_s = 0.457 \text{ in}^2/\text{ft}$ reqd.

USE #5 @ 8" o.c. A_s supplied $= 0.46 \text{ in}^2/\text{ft}$

CHECK BOND USING EQ 6.6 See Table 6.1 for α, β, γ, and λ

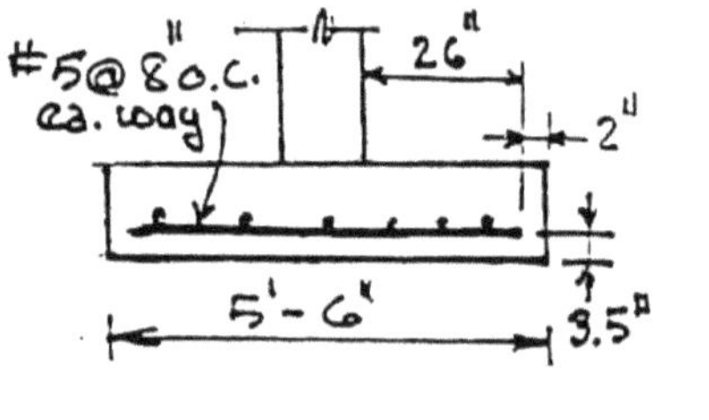

LENGTH AVAILABLE FOR ANCHORAGE

distance from face of wall to end of bar = 26"

FROM TABLE 6.1 : $\alpha = 1$ (bot bar), $\beta = 1$ (no coating) $\gamma = 0.8$, $\lambda = 1$

$K_{tr} = 0$ since no stirrups

$c = 3.5$ in. distance from ℄ of steel to out side surface

$$\frac{l_d}{d_b} = \frac{3 f_y \alpha \beta \gamma \lambda}{40\sqrt{f'_c}\left[\frac{c + K_{tr}}{d_b}\right]} = \frac{3 \times 60{,}000(1)(1)(0.8)(1)}{40\sqrt{3000}\,(2.5)} = 26.3$$

where $\frac{c + k_{tr}}{d_b} = \frac{3.5 + 0}{5/8} = 5.6$ but not to exceed 2.5

$l_d = 26.3 d_b = 26.3(5/8) = 16.4 \text{ in} < 26"$ ∴ OK

---

(2) **8.3** DESIGN A SQUARE FOOTING TO SUPPORT AN 18" SQUARE COLUMN. GROSS ALLOW SOIL PRESSURE = 10 K/FT^2

$f'_c = 3{,}000 \text{ lb/in}^2$, $f_y = 60{,}000 \text{ lb/in}^2$ $\gamma_s = 130 \text{ lb/ft}^3$

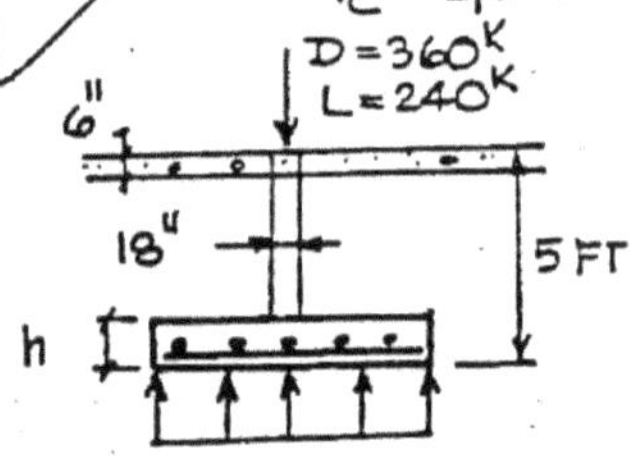

ESTIMATE DEPTH OF FOOTING $h = 2$ FT

NET ALLOW SOIL PRESSURE

$q_N = 10 - (2 \times .15 + 2.5 \times .13 + .5 \times .15)$
$= 9.3 \text{ K/FT}^2$

REQ'D FOOTING SIZE (BASE ON SERVICE Lds)

$A = \frac{P}{q_N} = \frac{600}{9.3} = 64.5 \text{ FT}^2$ USE 8'-2" SQUARE FOOTING $A = 66.7 \text{ FT}^2$

DESIGN :

$P_u = 1.4 \times 360^K + 1.7 \times 240^K = 912^K$

$q_u = \frac{P}{A} = \frac{912}{66.7} = 13.67 \text{ K/FT}^2$

8.3 continued

CHECK PUNCHING SHEAR CAPACITY (At d/2 OUT FROM COLUMN)

$d = h - (3''\text{ CLEAR COVER} + d_b)$ ASSUME $d_b = 1$ IN

$d = 24 - (4) = 20''$ $b_o = (18 + \frac{d}{2} \times 2)\,4 = 152$ IN

$$\phi V_c = (0.85)\,4\sqrt{f'_c}\,b_o d = (.85)\,4\frac{\sqrt{3000}}{1000}\,152\,(20) = \underline{581^K}$$

$$V_u = 912^K - \frac{38}{12} \times \frac{38}{12} \times 13.67 = 774.9^K > \phi V_c = 581^K$$

∴ NO GOOD INCREASE h

TRY h = 28 in ; THEN d = 24 IN

CHECK PUNCHING SHEAR $d = 24''$

$$\phi V_c = (0.85)\,4\frac{\sqrt{3{,}000}}{1{,}000}\,(4 \times 42)\,24 = 751^K$$

$$V_u = 912 - \left(\frac{42}{12}\right)^2 (13.67) = 744.5^K < \phi V_c = 751^K \ \therefore \text{ok.}$$

CHECK BEAM SHEAR ON SECTION d" OUT FROM COLUMN FACE

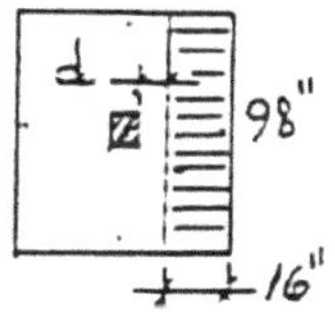

$$\phi V_c = (.85)\,2\sqrt{f'_c}\,b_w d = (.85)\,2\sqrt{3000}\,(98 \times 24)$$
$$= \underline{219 \text{ KIPS}}$$

$$V_u = \frac{16}{12} \times 8.16 \times 13.67 \text{ K/FT}^2 = 148.72^K$$

$\phi V_c > V_u$ ∴ OK

DESIGN FLEXURAL REINF. CRITICAL SECTION AT FACE OF COLUMN.

$$M_u = \frac{40}{12} \times 8.16 \times 13.67 \left(\frac{40}{12} \times \frac{1}{2}\right) = \underline{619.71 \text{ FT·K}}$$

COMPUTE A_s BY THE TRIAL METHOD

$M_u = \phi T (d - \frac{a}{2})$ GUESS $a = 2''$, $\phi = 0.9$, $T = A_s f_y$

$619.71 \times 12 = 0.9\, A_s\, 60 (24 - \frac{2}{2})$

$\underline{\underline{A_s = 7.25 \text{ IN}^2}}$

CHECK a , T = C
$7.25 \times 60 = 2.55\,(a \times 98)$
$a = 1.74 < 2''$ ∴ ok.

$$\text{Min. } A_s = \frac{200}{f_y}\,b_w d = \frac{200}{60{,}000}\,(98 \times 24) = 7.84 \text{ in}^2 \text{ controls}$$

Use 10 - #8 bars each way; $A_s = 7.86$ in^2

CHECK ANCHORAGE USING EQ 6.6. INCLUDE 0.8 reduction. SEE section (5) in Table 6.1

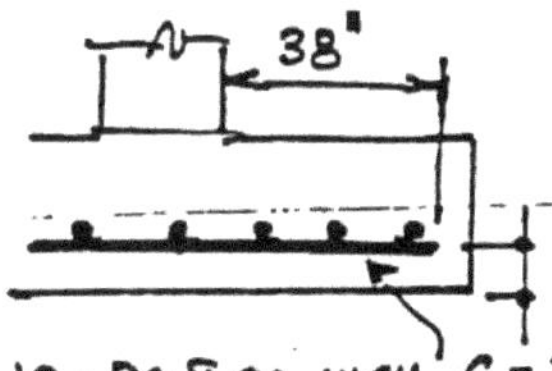

$$\frac{l_d}{d_b} = \frac{3 f_y\,\alpha\,\beta\,\gamma\,\lambda\,(0.8)}{40\sqrt{f'_c}\left[\frac{c + K_{tr}}{d_b}\right]_{\le 2.5}}$$

$$\frac{c + K_{tr}}{1} = 3.5 \quad (K_{tr} = 0,\ c = 3.5)\ \therefore \underline{\underline{\text{use } 2.5}}$$

$$l_d = \frac{3\,(60)(1)(1)(1)(1)\,0.8}{40\sqrt{3000}\,(2.5)} = 26.3 d_b = \underline{\underline{26.3''}} \quad (d_b = 1'')$$

$26.3'' = l_d < 38$ in ∴ ok

8.4 CONTINUED

CHECK ANCHORAGE USING EQ. 6.6: INCLUDE 0.8 REDUCTION DUE TO WIDE SPACING BETWEEN BARS. SEE SECTION (f) IN TABLE 6.1

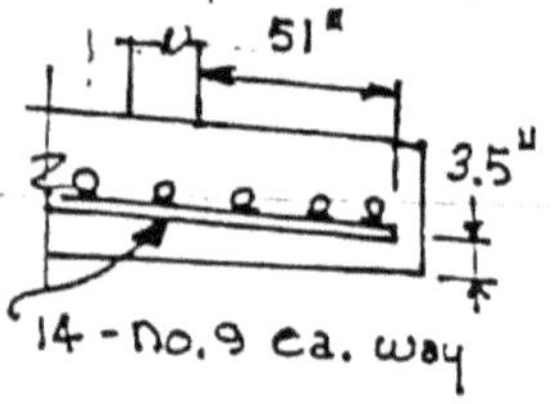

AVAILABLE ANCHORAGE LENGTH = 51"

$$\frac{l_d}{d_b} \quad \frac{3 f_y \alpha \beta \gamma \lambda (0.8)}{40\sqrt{f'_c}\left[\frac{c+K_{tr}}{d_b}\right]} \qquad \text{EQ. 6.6}$$

$$\frac{c+K_{tr}}{d_b} = \frac{3.5+0}{1.128} = 3.1 \text{ but not to exceed } 2.5$$

$$l_d/d_b = \frac{3(60{,}000)(1)(1)(1)(1)(0.8)}{40\sqrt{3000}\ (2.5)} = 26.3$$

$$l_d = 26.3 d_b = 26.3(1.128) = 29.7'' < 51'' \therefore \text{ OK}$$

---

(3) **8.5**

D = 300^K
L = 240^K

DESIGN A RECT. FTG TO SUPPORT SERVICE LOADS $D = 300^K$, $L = 240^K$; WIDTH NOT TO EXCEED 7 FT
$f'_c = 3.5$ K/in^2, $f_y = 50$ K/IN^2, $\gamma_s = 130$ lb/FT^3, $q_{ALLOW} = 8$ K/FT^2

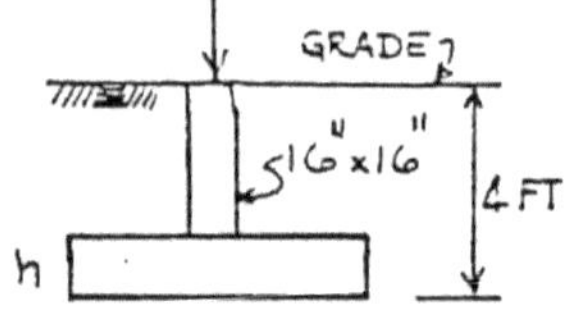

GUESS $h = 2$ FT $= 24$ IN

$q_{NET} = 8 - (.15\times2 + .13\times2) = 7.44$ K/FT^2

REQ'D AREA $= \frac{P}{q_{NET}} = \frac{540^K}{7.44} = 72.58$ FT^2

MAKE 7 FT × 10 FT - 6 IN, $A = 73.5$ FT^2

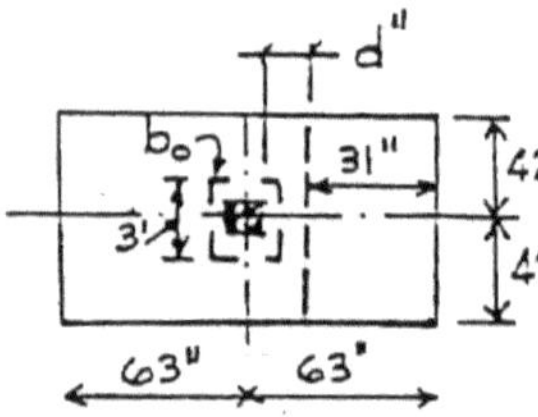

$P_u = 300^K(1.4) + 240^K(1.7) = 828^K$

$q_u = \frac{P_u}{A} = \frac{828}{73.5} = 11.265$ K/FT^2

CHECK PUNCHING SHEAR. CRITICAL PERIMETER d/2 OUT FROM FACE OF COLUMN. $d = h - 4'' = 20$ IN

$V_u = 828^K - (3'\times3')11.265 = 726.6^K$

$\phi V_c = (.85)4\sqrt{f'_c}\, b_o d = (.85)4\frac{\sqrt{3000}}{1000}(36\times4)(20) = 579.3^K$

$V_u > \phi V_c$ ∴ NO GOOD; OTHER TRIALS SHOW $h = 28''$ AND $d = 24''$ GIVES $\phi V_c = 772.4^K > V_u = 703^K$ ∴ OK

MAKE FTG. DEPTH 28"

CHECK BEAM SHEAR ON CRITICAL SECTION ⊥ TO LONG AXIS. GO d" OUT FROM FACE OF COLUMN

$V_u = \frac{31}{12}(7)\,11.265 = 203.71^K$

$\phi V_c = (.85)2\sqrt{f'_c}\, b_w d = (.85)2\frac{\sqrt{3500}}{1000}(84\times24) = 202.75^K < V_u = 203.71$

0.5% OVERSTRESS, SAY OK

8.5 CONT.

FLEXURAL STEEL, LONG DIRECTION (CRITICAL SECTION AT FACE OF COLUMN.

$$M_U = \frac{55}{12}(7)11.265\ K/FT^2\left(\frac{55}{12}\times\frac{1}{2}\right) = 828.25\ FT.K$$

COMPUTE A_S REQ'D FOR M_U. USE TRIAL METHOD

$$M_U = \phi A_s f_y\left(d - \frac{a}{2}\right) \quad \text{GUESS } a = 4''$$

$$828.25 \times 12 = 0.9 A_S 50\left(24 - \frac{4}{2}\right)$$

$A_S = 10.0\ IN^2$

VERIFY a; $T = C$

$10.0(50) = .85(3.5)84\,a$

$a = 2'' < 4''$ ∴ OK

ANALYSIS CONSERVATIVE

$$MIN\ A_S = \frac{200}{f_Y} b_w d = \frac{200}{50000}(84 \times 24) = 8.06\ IN < 10.0\ IN$$

USE 13-#8, SPACE EQUALLY ($A_S = 10.2\ in^2$)

FLEXURAL STEEL, SHORT DIRECTION

$$M_u = \frac{126}{12}\times\frac{34}{12}(11.265)\left(\frac{17}{12}\right) = 474.77\ FT\cdot K$$

COMPUTE A_S. USE THE TRIAL METHOD

$$M_U = \phi T\left(d - \frac{a}{2}\right) \quad \text{GUESS } a = 2''$$

$$12(474.77) = .9 A_s 50(24 - 1)$$

$A_S = 5.5\ IN^2$

CHECK a; $T = C$

$5.5(50) = .85(3.5)126\,a$

$a = .73'' < 2''$

∴ OK

$$MIN\ A_S = \frac{200}{f_Y} b_W d = \frac{200}{50{,}000}(126 \times 24) = 12.1\ in^2 \quad \text{CONTROLS}$$

DIVIDE STEEL IN SHORT DIRECTION INTO BANDS

$$A_S \text{ TO CENTER BAND OF WIDTH } b = 12.1\ \frac{2}{\left(\frac{10.5}{7}+1\right)} = 9.7\ IN^2$$

USE 13-#8

NOTE: BOND COMPUTATIONS NOT SHOWN

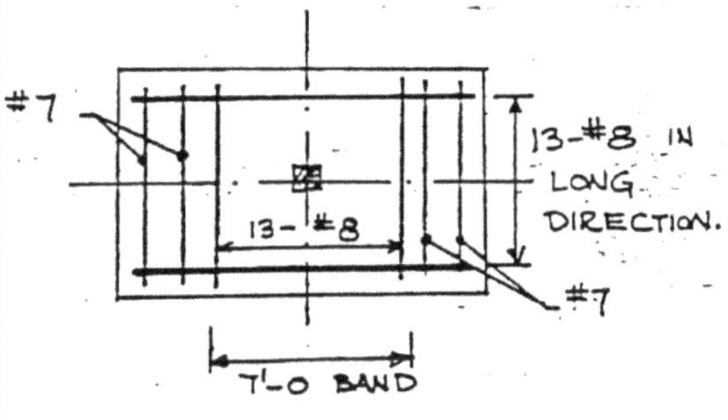

A_S TO OUTSIDE BAND = $12.1 - 9.7 = 2.4\ IN^2$

USE 4-#7, $A_S = 2.41\ IN^2$

Reinforced Concrete Design (2)

Homework_4-

5,2

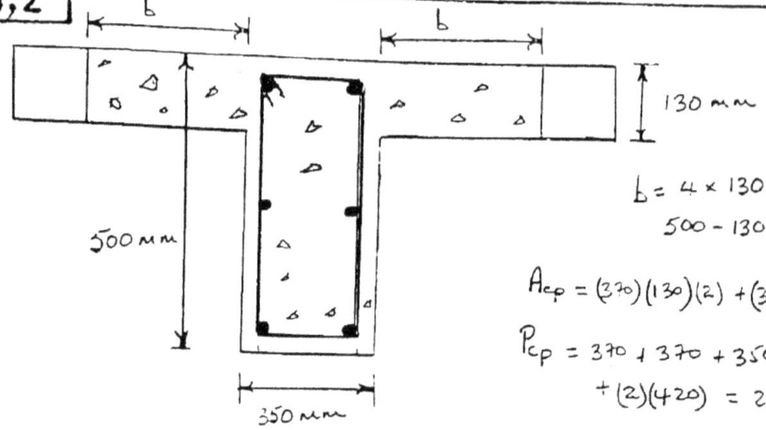

$b = 4 \times 130 = 520$ mm

$500 - 130 = 370 \Leftarrow$ CONTROLS

$A_{cp} = (370)(130)(2) + (350)(500) = 271,200\ mm^2$

$P_{cp} = 370 + 370 + 350 + (130)(2) + 350 + (2)(420) = 2540$ mm

$$\phi T_n = \frac{\phi \sqrt{f'_c}}{12}\left(\frac{A_{cp}^2}{P_{cp}}\right) = \frac{0.85\sqrt{30}}{12} \times \frac{(271,200)^2}{2540}$$

$= 11,234.2$ kN.mm $> T_u = 8,000$ kN.mm

∴ TORSION CAN be NEGLECTED

5.3

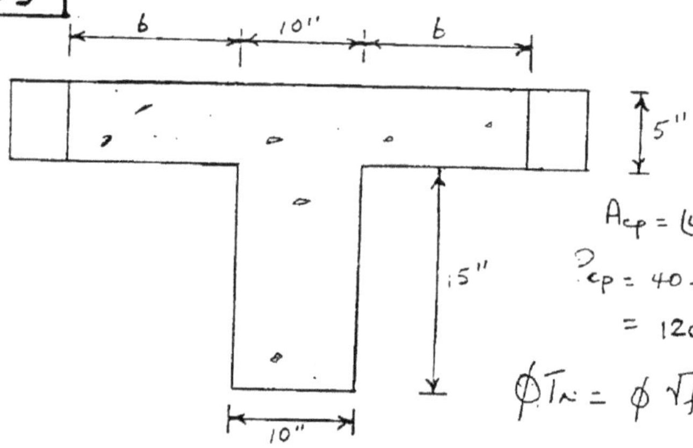

$b = (4)(5) = 20''$

$15'' \Leftarrow$ CONTROLS

$A_{cp} = (40)(5) + (15)(10) = 350\ in^2$

$P_{cp} = 40 + (2)(5) + 2(15) + 2(15) + 10 = 120$ in

$$\phi T_n = \phi \sqrt{f'_c}\ \frac{A_{cp}^2}{P_{cp}}$$

$$\phi T_n = 0.85 \times \sqrt{4000} \times \frac{(350)^2}{120} = 54.88\ \text{in-kips} < T_u = 120\ \text{in. kips}$$

∴ TORSION NEEDS To be CONSIDERED

5.4

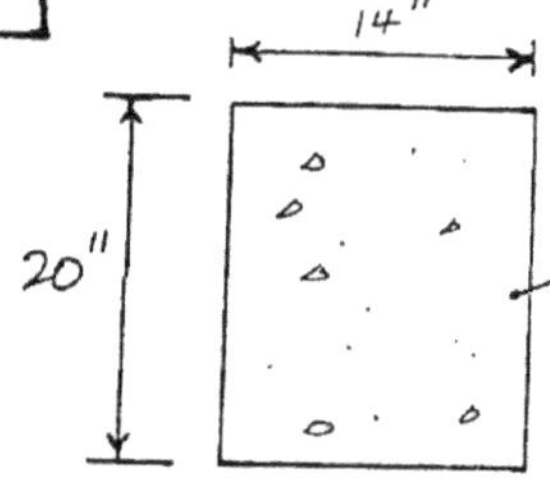

At THIS LOCATION THE SHEAR STRESSES FROM V_u & T_u are additive

5.6 CONT'D Since the computed Torque (65 ft.kips) is much larger than Tcr the beam will CRACK invalidating the relative stiffness used in the analysis

(b) A Torque equal to ϕT_{cr} is recommended by ACI 318-95 for these conditions – i.e. design for

$T_u = 0.85 \times 24.88$ OR $T_u = 21.15$ ft.kips

(c) The effective section is:

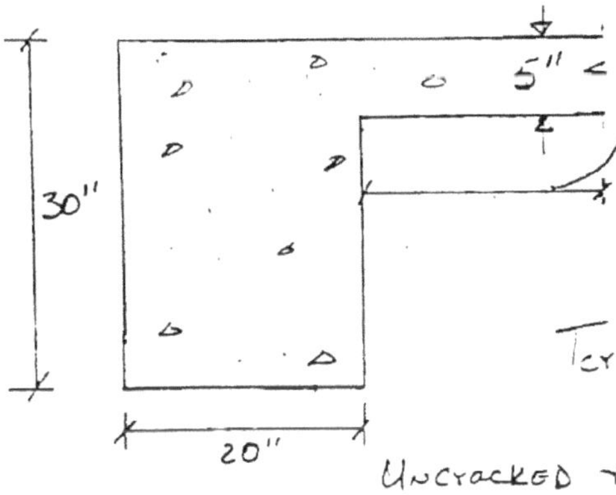

LESSER OF 30 − 5 = 25"
4 × 5" = 20" ← CONTROLS

$A_{cp} = 30 \times 20 + 20 \times 5 = 700\ in^2$

$P_{cp} = 30 + 20 + 25 + 20 + 5 - 40 = 140\ in$

$$T_{cr} = \frac{4 \times \sqrt{5000} \times (700)^2}{140\,(12)(1000)} = 82.5\ ft.kips$$

Uncracked response is anticipated therefore design for results of analysis

5.7 $V_u = 30$ kips, $M_u = 240$ ft-kips, $T_u = 100$ ft-kips

THE EQUATION THAT GOVERNS THE MINIMUM SECTION SIZE FOR COMBINED SHEAR AND TORSION IS:

$$\sqrt{\left(\frac{V_u}{b_w d}\right)^2 + \left(\frac{T_u P_h}{1.7 A_{oh}^2}\right)^2} \leq \phi\left(\frac{V_c}{b_w d} + 8\sqrt{f'_c}\right)$$

FOR THE WIDTH OF 14" WE HAVE

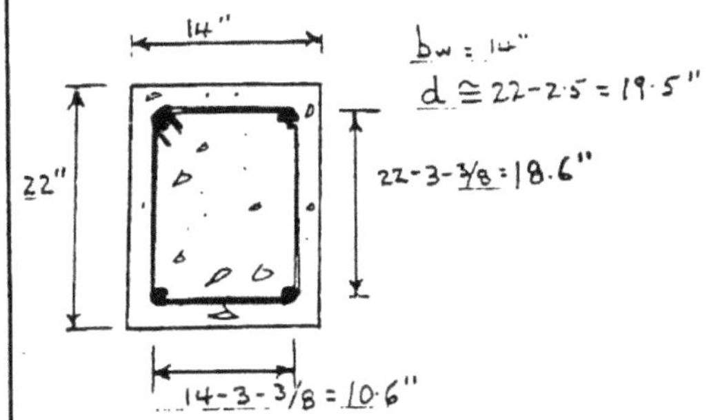

$b_w = 14"$

$d \cong 22 - 2.5 = 19.5"$

$A_{oh} = 18.6 \times 10.6 = 197.16\ in^2$

$P_h = (18.6 + 10.6)2 = 58.4\ in$

$$\sqrt{\left(\frac{30}{14 \times 19.5}\right)^2 + \left(\frac{100 \times 12 \times 58.4}{1.7 \times (197.16)^2}\right)^2}$$

$= 1.066$

5.7 CONT'D TAKING $V_c = 2\sqrt{f'_c}\, b_w d$ THE LIMIT IS $0.85\left(10\sqrt{3500}\right)\times\frac{1}{1000}$ $= 0.503$, SINCE $1.066 > 0.503$ THE SECTION IS NOT ADEQUATE.

THE SELECTION OF AN ACCEPTABLE WIDTH CAN BE CONVENIENTLY DONE BY TRIALS.-

TRY $b = 20''$

$$A_{oh} = 16.6 \times 18.6 = 308.76\ in^2$$

$$P_h = (18.6 + 16.6) \times 2 = 70.4\ in$$

THE RADICAL IS NOW

$$\sqrt{\left(\frac{30}{20\times 19.5}\right)^2 + \left(\frac{100\times 12\times 70.4}{1.7\times(308.76)^2}\right)^2} = 0.527$$

WHICH IS SLIGHTLY OVER THE 0.503 LIMIT, A 22 IN WIDTH WILL WORK

DESIGN OF REINFORCEMENT:

BENDING $A_s = \dfrac{M_u}{\phi F_y (d - a/2)}$ USE TRIAL METHOD, ASSUME $a = 2''$

$$A_s = \frac{240\times 12}{0.9\times 60(19.5-1)} = 2.88\ in^2 \quad (\text{WHERE } d \simeq 22 - 2.5 = 19.5'')$$

CHECK $a = \dfrac{A_s F_y}{b \times 0.85 f'_c} = \dfrac{2.88\times 60}{22\times 0.85\times 3.5} = 2.64''$

REFINE $A_s = \dfrac{240\times 12}{0.9\times 60\left(19.5 - \frac{2.64}{2}\right)} = 2.93\ in^2$

DETERMINE THE TRANSVERSE STEEL FOR SHEAR

$$V_c = 2\sqrt{f'_c}\, b_w d = 2\times\sqrt{3500}\times 22\times 19.5\times\frac{1}{1000} = 50.76\ \text{KIPS}$$

$$V_s = \frac{V_u}{\phi} - V_c = \frac{30}{0.85} - 50.76 = \text{NEGATIVE} \qquad \frac{A_v}{s} = 0$$

(MINIMUM IS APPLIED TO THE TOTAL WHICH INCLUDES THE TRANSVERSE STEEL FOR TORSION)

DETERMINE THE TRANSVERSE STEEL FOR TORSION

$$\frac{A_t}{s} = \frac{T_u}{2\phi F_{yv} A_o \cot\theta} = \frac{100\times 12}{2\times 0.85\times 60\times 294\times 1} = 0.04$$

WHERE $A_o = 0.85\times(18.6\times 18.6) = 294\ in^2$ AND $\theta = 45°$

TOTAL TRANSVERSE STEEL REQUIRED (TWO LEGS) $\left(\dfrac{A_v}{s}\right)_{TOTAL} = 2\times 0.04 = 0.08$

MINIMUM TRANSVERSE STEEL $= \dfrac{50 b_w}{F_{yv}} = \dfrac{50\times 22}{60{,}000} = 0.018 < 0.08$ DOES NOT CONTROL

5.10 $V_u = 160\ kN$, $M_u = 270\ kN.m$, $T_u = 120\ kN.m$

(a) THE EQUATION THAT GOVERNS MINIMUM SIZE IS

$$\sqrt{\left(\frac{V_u}{b_w d}\right)^2 + \left(\frac{T_u P_h}{1.7 A_{oh}^2}\right)^2} \leq \phi\left(\frac{V_c}{b_w d} + \frac{2}{3}\sqrt{f'c}\right)$$

WE HAVE:

$b_w = 350\ mm \qquad d = 500 - 65 = 435\ mm$

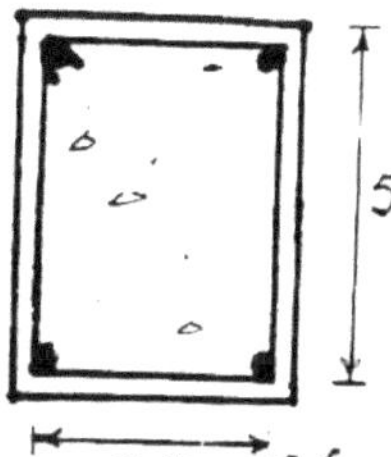

ASSUME #10 STIRRUP

$500 - 2(40+11.3) = 397.4\ mm$

$350 - 2(40+11.3) = 247.4\ mm$

$A_{oh} = 397.4 \times 247.4 = 98316.8\ mm^2$

$P_h = 2(397.4 + 247.4) = 1289.6\ mm$

THE LEFT SIDE OF THE INEQUALITY IS THEN

$$\sqrt{\left(\frac{160000}{350 \times 435}\right)^2 + \left(\frac{120 \times 10^6 \times 1289.6}{1.7 \times (98316.8)^2}\right)^2} = 9.48\ MPa$$

TAKING $\dfrac{V_c}{b_w d} = \dfrac{\sqrt{f'c}}{6}$ THE RIGHT SIDE LIMIT IS

$$0.85 \times \frac{5}{6}\sqrt{f'c} = 0.85 \times \frac{5}{6}\sqrt{21} = 3.24\ MPa$$

SINCE $9.48 > 3.24$ <u>THE SECTION IS TOO SMALL</u>

(b)

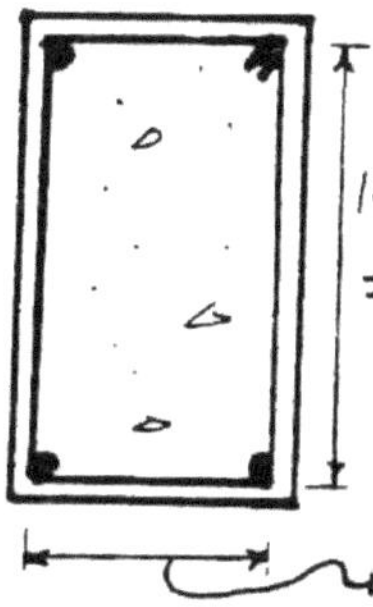

ASSUME #10 STIRRUP

$1000 - 2(40+11.3) = 897.4\ mm$

$600 - 2(40+11.3) = 497.4\ mm$

WE HAVE:

$b_w = 600\ mm$

$d = 1000 - 65 = 935\ mm$

$A_{oh} = 497.4 \times 897.4 = 446,366.8\ mm^2$

$P_h = 2(897.4 + 497.4) = 2789.6\ mm$

$A_{cp} = 600 \times 1000 = 600,000\ mm^2$

5.10 CONT'D DESIGN OF REINFORCEMENT

BENDING: $A_s = \frac{M_u}{\phi f_y (d - \frac{a}{2})}$ ASSUME $a = 60$ mm

$$A_s = \frac{270 \times 10^6 \text{ N.mm}}{0.9 \times 400 \times 10^6 \text{ N/mm}^2 \times 10^{-6} \text{ m}^2/\text{mm}^2 \left(935 - \frac{60}{2}\right)} = 829 \text{ mm}^2$$

CHECK $a = \frac{A_s f_y}{b \cdot 0.85 f'c} = \frac{829 \text{ mm}^2 \times 400 \text{ N/mm}^2}{600 \text{ mm} \times 0.85 \times 21 \text{ N/mm}^2} = 30.96$ mm

REFINE $A_s = \frac{270 \times 10^6}{0.9 \times 400 \times \left(935 - \frac{30.96}{2}\right)} = 815.64 \text{ mm}^2$

TRANSVERSE STEEL FOR TORSION

$$\frac{A_t}{S} = \frac{T_u}{2 \phi f_{yv} A_o \cot\theta} = \frac{120 \times 10^6 \text{ N.mm}}{2 \times 0.85 \times 400 \text{ N/mm}^2 \times 379412 \text{ mm}^2 \times 1} = 0.465 \text{ mm}$$

TOTAL TRANSVERSE STEEL REQUIRED

$$= \left(\frac{A_v}{S}\right)_{TOTAL} = 2 \times 0.465 = 0.93 \text{ mm}$$

$$\left(\frac{A_v}{S}\right)_{MIN} = \frac{0.35 b_w}{f_{yv}} = \frac{0.35 \times 600}{400} = 0.525 \text{ mm}$$ ← DOES NOT CONTROL

LONGITUDINAL TORSION STEEL

$$A_L = \frac{A_t}{S} P_h \left(\frac{f_{yv}}{f_{yL}}\right) \cot^2\theta = 0.465 \times 2789.6 \times 1 \times 1 = 1297.2 \text{ mm}^2$$

$$(A_L)_{MIN} = \frac{5\sqrt{f'c} A_{cp}}{12 f_{yL}} - \frac{A_t}{S} P_h \frac{f_{yv}}{f_{yL}}$$

$$(A_L)_{MIN} = \frac{5 \times \sqrt{21} \times 600{,}000}{12 \times 400} - 0.465 \times 2789.6 \times 1 = 1566.9 \text{ mm}^2$$

CONTROLS

Reinforced Concrete Design (2)

Exam -1- / Solutions

(1) (a) Using the alignment charts, calculate the effective length factor for column AB of the unbraced frame shown in the figure. Consider only bending in the plane of the frame.

(b) Compute the slenderness ratio of column AB. Is it a short or slender column?

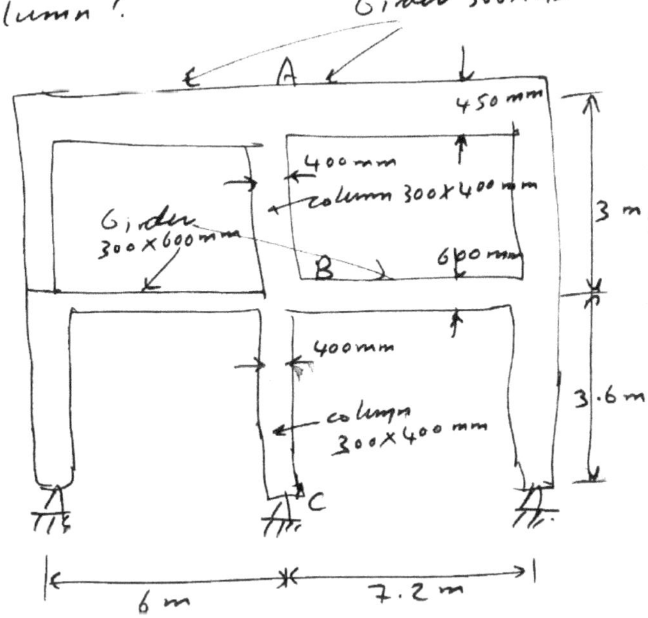

Solution:

(a) Top Girder:

$I_{gross} = \frac{1}{12}(300)(450)^3$

$= 2.278125 \times 10^9 \ mm^4.$

$I = 0.35 \ I_{gross}$

$= 0.35 \ (2.278125 \times 10^9)$

$= 0.79734 \times 10^9 \ mm^4.$

Middle Girder:

$I_{gross} = \frac{1}{12}(300)(600)^3$

$= 5.4 \times 10^9 \ mm^4.$

$I = 0.35 \ I_{gross}$

$= 0.35 \ (5.4 \times 10^9)$

$= 1.89 \times 10^9 \ mm^4.$

Columns AB & BC:

$I_{gross} = \frac{1}{12}(300)(400)^3 = 1.6 \times 10^9 \ mm^4.$

$I = 0.7 \ I_{gross}$

$= 0.7 \ (1.6 \times 10^9) = 1.12 \times 10^9 \ mm^4.$

12

For column AB:

Top: $\Sigma(I_c/L_c) = \dfrac{1.12\times10^9}{3\times1000} = 0.3733\times10^6\ mm^3.$

$\Sigma(I_g/L_g) = \dfrac{0.79934\times10^9}{6\times1000} + \dfrac{0.79734\times10^9}{7.2\times1000}$

$= 0.24363\times10^6\ mm^3.$

$\psi_A = \dfrac{\Sigma(I_c/L_c)}{\Sigma(I_g/L_g)} = \dfrac{0.3733\times10^6}{0.24363\times10^6} = \boxed{1.53}$

Bottom: $\Sigma(I_c/L_c) = \dfrac{1.12\times10^9}{3\times1000} + \dfrac{1.12\times10^9}{3.6\times1000}$

$= 0.68444\times10^6\ mm^3.$

$\Sigma(I_g/L_g) = \dfrac{1.89\times10^9}{6\times1000} + \dfrac{1.89\times10^9}{7.2\times1000}$

$= 0.5775\times10^6\ mm^3.$

$\psi_B = \dfrac{\Sigma(I_c/L_c)}{\Sigma(I_g/L_g)} = \dfrac{0.68444\times10^6}{0.5775\times10^6} = \boxed{1.185}$

Reading from the alignment chart, we get $\boxed{k = 1.5}$.

(b) $l_u = (3\times1000) - \dfrac{450}{2} - \dfrac{600}{2} = 2475\ mm.$

$r = 0.3h = 0.3(400) = 120\ mm.$

$\dfrac{kl_u}{r} = \dfrac{(1.5)(2475)}{120} = 30.94 > 22$

$\Rightarrow$ the column is long (slender).

(2) The 300X500mm tied column shown in the figure is to be used to support the following : $P_D = 445$ kN , $P_L = 490$ kN, $M_D = 81$ kN.m, and $M_L = 108$ kN.m . If $f'_c = 28$ MPa (4 ksi) and $f_y = 400$ MPa (60 ksi), select the reinforcing bars to be used in two end faces by using the interaction curves.

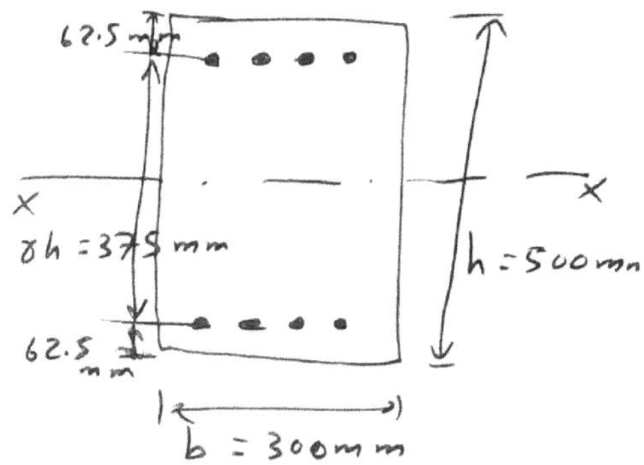

Solution :

$P_u = 1.4 P_D + 1.7 P_L$

$= 1.4(445) + 1.7(490)$

$= 1456$ kN.

$M_u = 1.4 M_D + 1.7 M_L$

$= 1.4(81) + 1.7(108)$

$= 297$ kN.m .

$$e = \frac{M_u}{P_u} = \frac{297 \times 1000}{1456} = 204 \text{ mm.}$$

Set $\phi P_n = P_u = 1456$ kN.

$A_g = 300 \times 500 = 150{,}000 \text{ mm}^2$.

$$\frac{\phi P_n}{A_g} = \frac{1456}{150{,}000 \times 10^{-6}} \times 10^{-3} = 9.71 \text{ MPa} = \frac{9.71}{6.895} = 1.408 \text{ ksi.}$$

$$\frac{e}{h} = \frac{204}{500} = 0.408 .$$

$$\frac{\phi P_n}{A_g}\frac{e}{h} = (1.408)(0.408) = 0.574$$

$$\gamma = \frac{\gamma h}{h} = \frac{375}{500} = 0.75$$

Read from the interaction curve, $\rho_g = 0.020 = 2\%$.

$A_s = \rho_g A_g = (0.020)(150{,}000) = 3000 \text{ mm}^2$.

Use $8 \phi 22$ mm ($A_{s,supplied} = 8\frac{\pi(22)^2}{4} = 3041 \text{ mm}^2$).

(3) Determine the design capacity P_n of the short tied column shown in the figure which is subjected to biaxial bending. Use $f'_c = 28$ MPa (4 ksi), $f_y = 400$ MPa (60 ksi), $e_x = 400$ mm and $e_y = 200$ mm. (Hint: Use the Bresler equation).

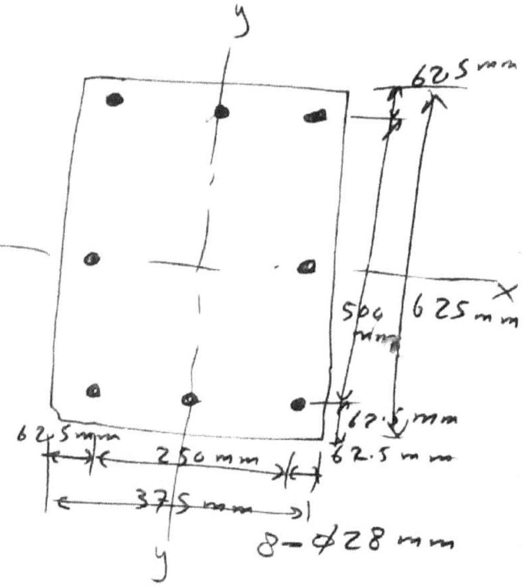

Solution:

Bending about the x-x axis:

$$\gamma = \frac{500}{625} = 0.8$$

$$\frac{e_x}{h} = \frac{400}{625} = 0.64$$

$$A_g = 625 \times 375 = 234{,}375 \text{ mm}^2.$$

$$A_{st} = 8\,\frac{\pi(28)^2}{4} = 4926 \text{ mm}^2.$$

$$\rho_g = \frac{A_{st}}{A_g} = \frac{4926}{234{,}375} = 0.0210$$

From the interaction curve, read $\dfrac{\phi P_{nx}}{A_g}\dfrac{e}{h} = 0.523$ ksi

$= 0.523 \times 6.895 = 3.61$ MPa

$$\therefore P_{nx} = \frac{(3.61 \times 1000)(234{,}375 \times 10^{-6})}{(0.7)(0.64)} = 1889 \text{ kN}.$$

Bending about the y-y axis:

$$\gamma = \frac{250}{375} = 0.667$$

$$\frac{e_y}{h} = \frac{200}{375} = 0.533, \quad \rho_g = 0.0210$$

From the interaction curve, read $\dfrac{\phi P_{ny}}{A_g}\dfrac{e}{h} = 0.487$ ksi

$= 0.487 \times 6.895$

$= 3.36$ MPa.

$$P_{ny} = \frac{(3.36 \times 1000) \times 234{,}375 \times 10^{-6}}{(0.7)(0.533)} = 2111 \text{ kN}.$$

Pure axial load:

$$P_o = 0.85 f_c'(A_g - A_{st}) + f_y A_{st}$$

$$= 0.85(28 \times 1000)\left[234{,}375 - 4926\right] \times 10^{-6} + (400 \times 1000)(4926 \times 10^{-6})$$

$$= 7431 \text{ kN}.$$

Use the Bresler equation:

$$\frac{1}{P_n} = \frac{1}{P_{nx}} + \frac{1}{P_{ny}} - \frac{1}{P_o}$$

$$= \frac{1}{1889} + \frac{1}{2111} - \frac{1}{7431}$$

$$\therefore \boxed{P_n = 1151 \text{ kN}.}$$

Reinforced Concrete Design (2)

Exam -1- / Solutions

1

(1) It is assumed that the column shown in the figure has a strain on its compression edge equal to -0.003 and has a tensile strain of $+0.002$ on its other edge. Determine the values of P_n and M_n which cause this strain distribution if $f_y = 400$ MPa and $f_c' = 28$ MPa.

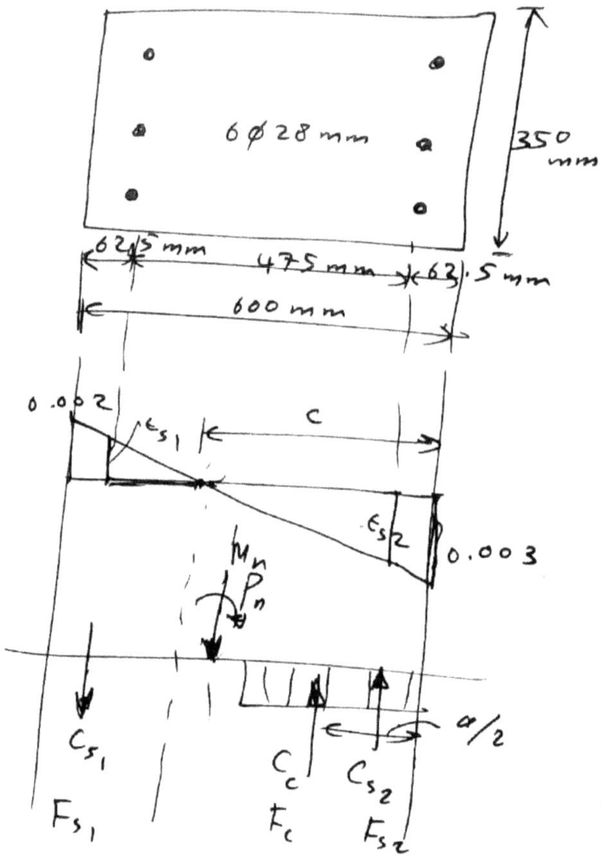

Solution:

From similar triangles:

$$\frac{c}{0.003} = \frac{600 - c}{0.002}$$

$$0.002c = 1.8 - 0.003c$$

$$c = 360 \text{ mm}.$$

$$\Rightarrow 600 - 360 = 240 \text{ mm}.$$

$$\frac{\epsilon_{s1}}{240 - 62.5} = \frac{0.002}{240}$$

$$\Rightarrow \epsilon_{s1} = 0.0014792$$

$$\frac{\epsilon_{s2}}{360 - 62.5} = \frac{0.003}{360}$$

$$\epsilon_{s2} = 0.002479$$

$$\epsilon_y = \frac{f_y}{E_s} = \frac{400}{200{,}000} = 0.002$$

$\epsilon_{s1} = 0.0014792 < \epsilon_y = 0.002 \Rightarrow$ tension steel did not yield.

$\epsilon_{s2} = 0.002479 > \epsilon_y = 0.002 \Rightarrow$ compression steel yields.

2

$a = 0.85c = 0.85(360) = 306 \text{ mm}.$

$F_c = C_c = 0.85 f_c' a b = 0.85(28 \times 1000)\left(\frac{306}{1000}\right)\left(\frac{350}{1000}\right) = 2549 \text{ kN}.$

$f_{s_1} = E_s \epsilon_{s_1} = (200{,}000)(0.0014792) = 295.84 \text{ MPa}$

$f_{s_2} = E_s \epsilon_y = (200{,}000)(0.002) = 400 \text{ MPa} = f_y.$

$F_{s_1} = A_{st} \cdot f_{s_1} = \left[\frac{3\pi(28)^2}{4} \times 10^{-6}\right](295.84 \times 10^3) = 546.5 \text{ kN}$

$F_{s_2} = A_{st}(f_{s_2} - 0.85 f_c')$

$= \left[\frac{3\pi(28)^2}{4} \times 10^{-6}\right]\left[400 - 0.85(28)\right] \times 10^3 = 694.9 \text{ kN}.$

$+\uparrow \Sigma F_y = 0: \quad -P_n - F_{s_1} + F_c + F_{s_2} = 0$

$P_n = F_c + F_{s_2} - F_{s_1} = 2549 + 694.9 - 546.5$

$P_n = \boxed{2697.4 \text{ kN}.}$

$+\circlearrowleft \Sigma M_{\text{tension steel}} = 0:$

$-P_n(300 - 62.5) - M_n + C_{s_2}(600 - 62.5 - 62.5)$ (C_{s_2} ← F_{s_2})

$+ F_c\left(600 - 62.5 - \frac{306}{2}\right) = 0$

$M_n = -2697.4(237.5) + (694.9)(475) + 2549(384.5)$

$= 669535 \text{ kN.mm}$

$M_n = \boxed{669.5 \text{ kN.m}}$

(2) The tied column shown in the figure has been approximately sized to the dimensions 300 mm × 375 mm. It is to be used in a frame braced against sidesway. The column is bent in single curvature about its y-axis and has a clear height $l_u = 4.8$ m. If $k = 0.83$, $f_y = 400$ MPa (60 ksi), and $f_c' = 28$ MPa (4 ksi), determine the reinforcement required. Consider only bending in the plane of the frame. Note also that the dead axial load $P_D = 134$ kN. [3

Solution:

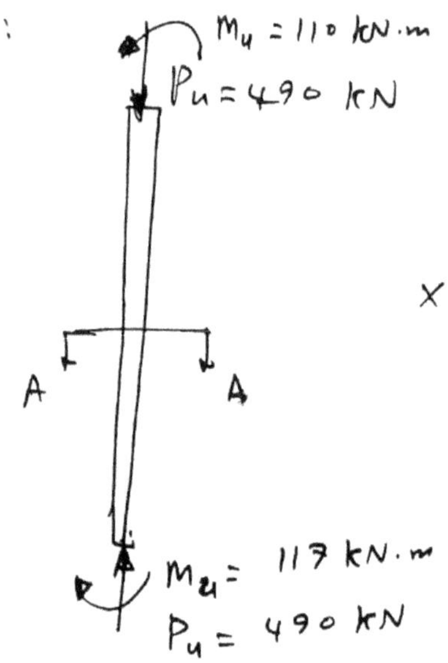

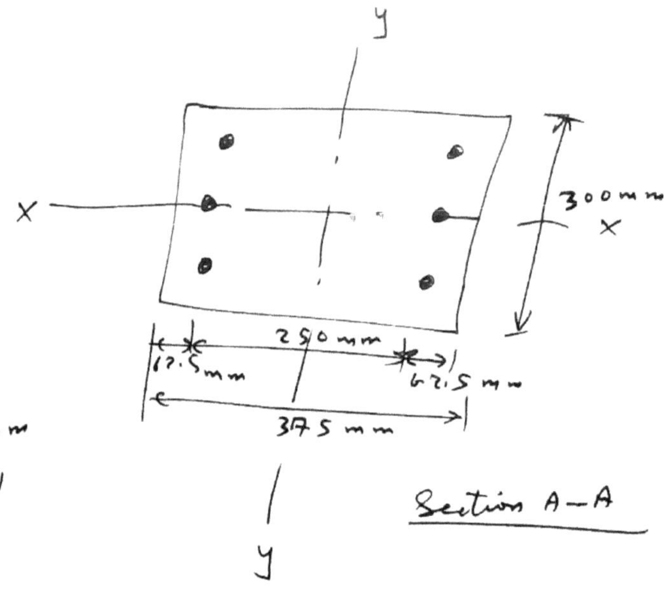

Section A–A

$$\frac{M_1}{M_2} = \frac{110}{117} = 0.94 > -0.50$$

$$34 - 12\frac{M_1}{M_2} = 34 - 12(0.94) = 22.7$$

$$r = 0.3h = 0.3 \times 375 = 112.5 \text{ mm}$$

$$\frac{K l_u}{r} = \frac{(0.83)(4.8) \times 1000}{112.5} = 35.4 > 22.7$$

$\Rightarrow$ the column is long (slender)

So, we need moment magnification. (4

$E_c = 5000\sqrt{f_c'} = 5000\sqrt{28} = 26{,}458$ MPa.

$I_g = \frac{1}{12}(300)(375)^3 = 1318.36 \times 10^6$ mm^4.

$$\beta_d = \frac{1.4 P_D}{P_u} = \frac{1.4 \times 134}{490} = 0.383.$$

$$EI = \frac{E_c I_g / 2.5}{1+\beta_d} = \frac{(26{,}458 \times 10^3 \times 1318.36 \times 10^{6} \times 10^{-12})/2.5}{1+0.383}$$

$$= 10{,}123 \text{ kN}\cdot\text{m}^2.$$

$$P_c = \frac{\pi^2 EI}{(k\ell_u)^2} = \frac{\pi^2 (10{,}123)}{[0.83(4.8)]^2} = 6295 \text{ kN}.$$

$$C_m = 0.6 + 0.4\frac{M_1}{M_2} = 0.6 + 0.4\left(\frac{110}{117}\right) = 0.976.$$

$$\delta_{ns} = \frac{C_m}{1 - \frac{P_u}{0.75 P_c}} = \frac{0.976}{1 - \frac{490}{(0.7)(6295)}} = 1.098 \approx 1.10.$$

$$M_u = M_c = \delta_{ns} M_2 = (1.098)(117) = 128.5 \text{ kN}\cdot\text{m}.$$

Check minimum $M_{2(min)}$. $= (15 + 0.03h)\frac{P_u}{1000} = 12.86$ kN·m.

$$\frac{\phi P_n}{A_g} = \frac{P_u}{A_g} = \frac{490}{300 \times 375 \times 10^{-6}} \times 10^{-3} = 4.356 \text{ MPa}$$

$$= 0.632 \text{ ksi}.$$

$$e = \frac{M_u}{P_u} = \frac{128.5 \times 1000}{490} = 262 \text{ mm}.$$

15

$\therefore \frac{\phi P_n}{A_g} \cdot \frac{e}{h} = (\sout{0.245}\ 0.632)\left(\frac{262}{375}\right) = 0.441$ ksi.

$\gamma = \frac{250}{375} = 0.667$.

By interpolation from the interaction curves,

$\rho_g = 0.0158$.

$A_{st} = \rho_g A_g = (0.0158)(300 \times 375) = 1778\ mm^2$.

Use $4\phi25$ mm

$A_{s(provided)} = \frac{4\pi(25)^2}{4} = 1963\ mm^2$.

Design of Ties: Use $\phi10$mm

smaller of $\left\{\begin{array}{l} 16 \times 25 = 400mm \\ 300\ mm \\ 48 \times 10 = 480mm \end{array}\right\}$ use $\phi10$mm @ 300mm c/c.

(1)

Reinforced Concrete Design (2)

Exam-2 - / Solutions

Consider a <u>rectangular</u> footing for a 450 mm square edge column with a dead load of 820 kN and a live load of 670 kN. Make the length of the long side equal to twice the width of the short side. Take $f_y = 400$ MPa, $f_c' = 28$ MPa, $\gamma_s = 2000\ /m^3$ and the allowable soil pressure on the base of the footing is $q_a = 200\ kN/m^2$. Assume the base of the footing is 1.5 m below grade.

(a) Establish the dimensions of the footing.

(b) Check punching shear (two-way shear) and beam shear (one-way shear).

(c) Determine the flexural reinforcement in the long direction.

(d) Determine the flexural reinforcement in the short direction.

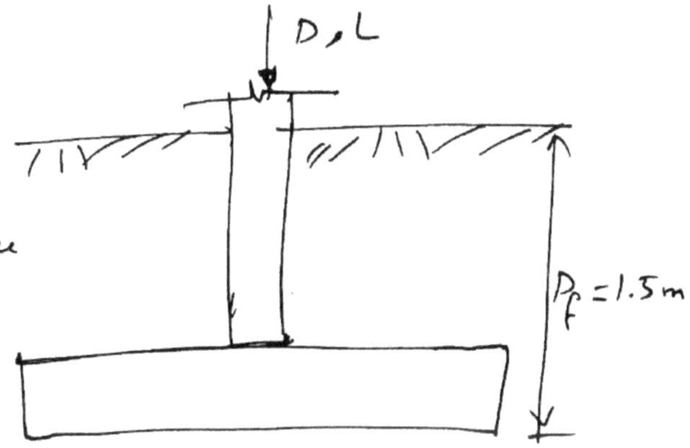

Solution:

(a) Establish the dimensions of the base using the total service load

$D + L = 820 + 670 = 1490$ kN.

(2)

Net allowable soil pressure

$$q_n = 200 - \left(\frac{1.5 \times 2000 \times 9.81}{1000}\right) = 171 \text{ kN/m}^2.$$

$$\text{Required Area} = \frac{D+L}{q_n} = \frac{1490}{171} = 8.71 \text{ m}^2.$$

Length = L = 2w , w = width

Area = 2w×w = 8.71 ⟹ $\boxed{w = 2.1 \text{ m}}$

Use a rectangular footing of 2.1× 4.2 m.

Area = 2.1×4.2 = 8.82 m².

(b) Try a 600 mm depth (Use h = 600 mm).

$d = 600 - 75 - 25 = 500$ mm.

$$P_u = 1.4D + 1.7L = (1.4 \times 820) + (1.7 \times 670) = 2287 \text{ kN}.$$

$$P_u = \frac{P_u}{\text{Area}} = \frac{2287}{8.82} = 259 \text{ kN/m}^2.$$

<u>Check punching shear:</u>

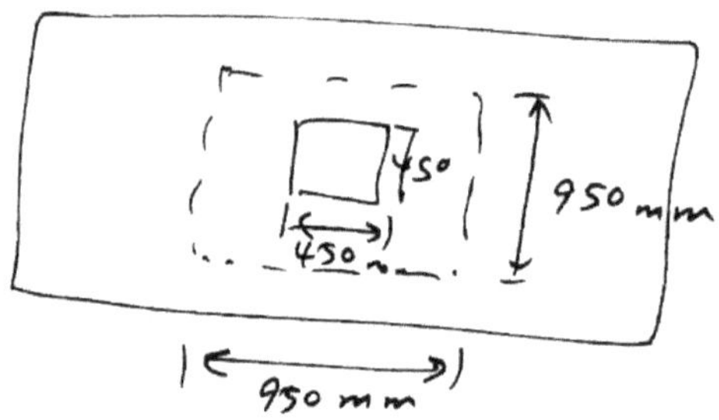

$V_u = P_u$ - soil pressure

$= 2287 - 259(0.95)(0.95)$

$= 2053$ kN.

$b_o = 4(950) = 3800$ mm.

$$\phi V_c = \text{smaller of} \begin{cases} \phi \frac{1}{3}\sqrt{f_c'}\, b_o d = 0.85\left(\frac{1}{3}\right)\sqrt{28}\left(\frac{3800}{1000}\right)(500) = 2849 \text{ kN}. \\ \phi\left(\frac{\alpha_s d}{b_o} + 2\right)\frac{1}{12}\sqrt{f_c'}\, b_o d = 0.85\left(\frac{40(500)}{3800} + 2\right)\frac{1}{12}\sqrt{28}\left(\frac{3800}{1000}\right)(500) \\ = 5172 \text{ kN}. \end{cases}$$

(5)

$\therefore \phi V_c = 2849 \text{ kN}$.

$V_u = 2053 \text{ kN} < \phi V_c = 2849 \text{ kN} \Rightarrow$ o.k.

Check beam shear in the long direction:

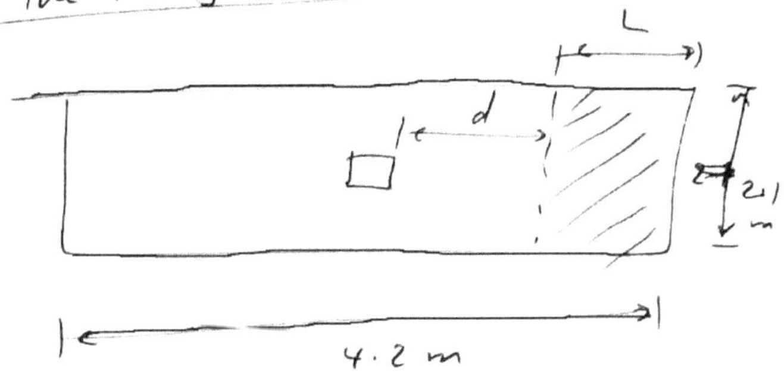

$$L = \frac{4200 - 450 - 2(500)}{2}$$

$= 1375$ mm.

$V_u = P_u \text{ (area)}$

$= 259 \times \left(\frac{1375}{1000}\right)(2.1) = 748 \text{ kN}$.

$\phi V_c = \phi \frac{1}{6} \sqrt{f_c'}\, b_w d = 0.85 \left(\frac{1}{6}\right) \sqrt{28}\,(2.1)(500) = 787 \text{ kN}$.

$V_u = 748 \text{ kN} < \phi V_c = 787 \text{ kN} \Rightarrow$ o.k.

Check beam shear in the short direction:

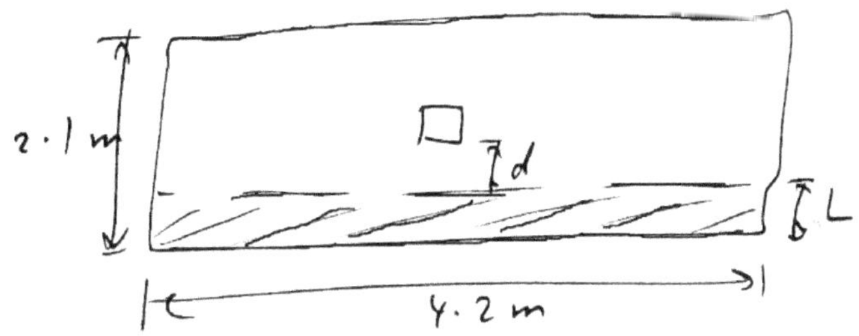

$$L = \frac{2100 - 450 - 2(500)}{2}$$

$= 325$ mm.

$V_u = 259 \times \frac{325}{1000} \times 4.2$

$= 354 \text{ kN}$.

$\phi V_c = \phi \frac{1}{6} \sqrt{f_c'}\, b_w d = (0.85)\left(\frac{1}{6}\right) \sqrt{28}\,(4.2)(500)$

$= 1574 \text{ kN}$.

$V_u = 354 \text{ kN} < \phi V_c = 1574 \text{ kN} \Rightarrow$ o.k.

(4)

(c) Determine reinforcement in the long direction:

$L = \frac{4200 - 450}{2} = 1875\ mm.$

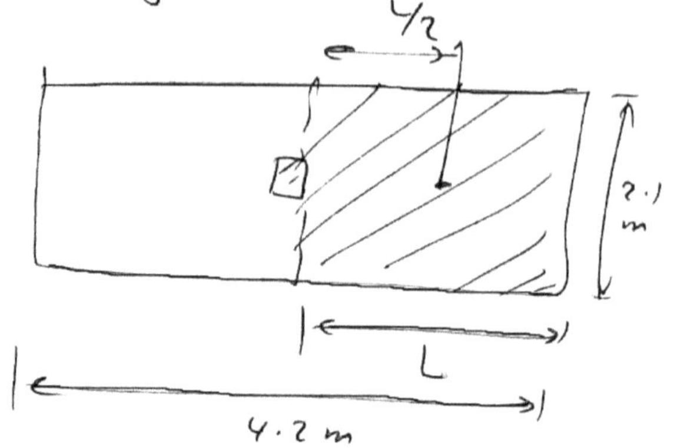

$M_u = P_u (area) \cdot \frac{L}{2}$

$= 259 \times \left(\frac{1875}{1000}\right)(2.1)\left(\frac{1}{2}\right)\left(\frac{1875}{1000}\right)$

$= 956\ kN \cdot m.$

Guess $a = 50\ mm.$

$M_u = \phi f_y A_s (d - a/2)$

$956 = 0.9(400 \times 1000)(A_s)\left(\frac{500}{1000} - \frac{50}{2(1000)}\right)$

$\Rightarrow A_s = 0.005591\ m^2$

$= 5591\ mm^2.$

Check a : $C = T$

$0.85 f'_c a b = f_y A_s$

$0.85(28)a(2.1 \times 1000) = 400 \times 5591$

$\Rightarrow a = 44.7\ mm$ close to 50 mm.

$A_{s(min)} = \frac{1.4 b_w d}{f_y} = \frac{1.4 \times 2.1 \times 1000 \times 500}{400} = 3675\ mm^2.$

Use $A_s = 5591\ mm^2.$

Use $10\phi28\ mm \Rightarrow A_{s(used)} = \frac{10\pi(28)^2}{4} = 6158\ mm^2.$

Check development length: (5)

$\frac{2100}{10} = 210$ mm spacing between bars.

$c = \begin{cases} \frac{210}{2} = 105 \text{ mm} \\ 75 + \frac{1}{2}(28) = 89 \text{ mm} \end{cases}$ use $c = 89$ mm.

$\frac{c + K_{tr}}{d_b} = \frac{89 + 0}{28} = 3.2 \not< 2.5$ Use 2.5.

$\alpha = 1$, $\beta = 1$, $\gamma = 1$, $\lambda = 1$.

$\frac{l_d}{d_b} = \frac{15 f_y \alpha \beta \gamma \lambda}{16 \sqrt{f_c'} \left(\frac{c + K_{tr}}{d_b}\right)} = \frac{15 \times 400 \times 1 \times 1 \times 1 \times 1}{16 \sqrt{28} \times 2.5} = 28.35$

$\therefore l_d = 28.35 \times 28 = 794$ mm.

Available anchorage length $= 1875 - 50 = 1825$ mm > 794 mm

o.k

(d) Determine reinforcement in the short direction:

$L = \frac{2100 - 450}{2}$

$= 825$ mm.

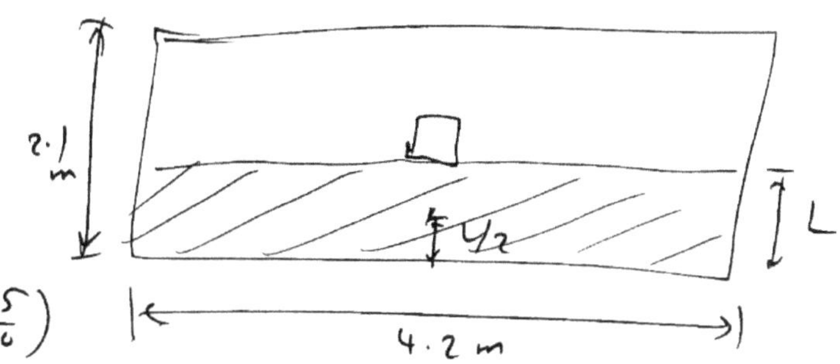

$M_u = 259 \times \frac{825}{1000} \times 4.2 \times \frac{1}{2}\left(\frac{825}{1000}\right)$

$= 370$ kN.m.

$M_u = \phi f_y A_s \left(d - \frac{a}{2}\right)$, Guess $a = 15$ mm.

$370 = 0.9(400 \times 1000) A_s \left(\frac{500}{1000} - \frac{1}{2}\frac{15}{1000}\right)$

$A_s = 0.002087 \text{ m}^2 = 2087 \text{ mm}^2$.

(6)

Check a:

$$C = T$$

$$0.85 f_c' a b = f_y A_s$$

$$0.85 (28)\, a\, (4.2 \times 1000) = 400 \times 2087$$

$$\Rightarrow a = 8.35 \text{ mm close to } 15 \text{ mm.}$$

$$A_{s(min)} = \frac{1.4 b_w d}{f_y} = \frac{1.4 \times 4.2 \times 1000 \times 500}{400} = 7350 \text{ mm}^2.$$

Use $A_s = 7350 \text{ mm}^2$.

Use 12 ϕ 28 mm $\Rightarrow$ $A_{s(used)} = \frac{12\pi (28)^2}{4} = 7389 \text{ mm}^2$.

Divide steel in the short direction between a center band:

$$\beta = \frac{4.2}{2.1} = 2.$$

$$\text{No. of bars in band} = \left(\frac{2}{1+\beta}\right)(12) = \left(\frac{2}{3}\right)(12) = 8.$$

Use 8 bars in the center band and 4 bars outside the band.

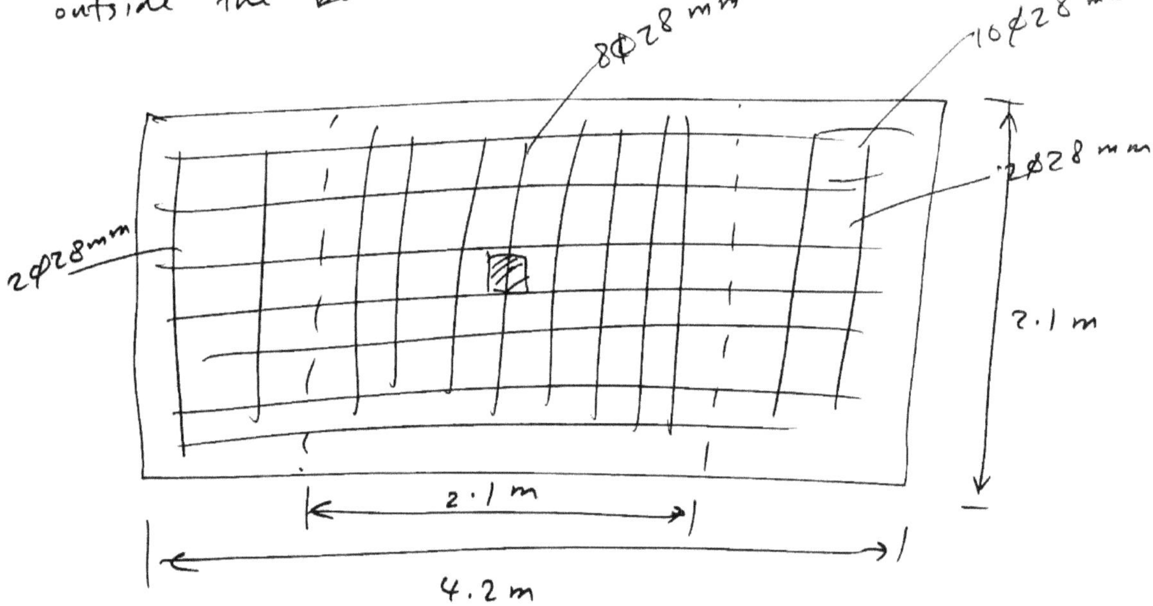

(7)

Check development length :

$$\frac{2100}{8} = 262.5 \text{ mm} \quad \text{spacing of bars in band.}$$

$$c = \begin{cases} \frac{262.5}{2} = 131 \text{ mm} \\ 75 + \frac{1}{2}(28) = 89 \text{ mm} \end{cases} \quad \text{use } c = 89 \text{ mm.}$$

$$\frac{c + K_{tr}}{d_b} = \frac{89 + 0}{28} = 3.2 \not< 2.5 \Rightarrow \text{use } \underline{\underline{2.5}}.$$

$$\alpha = \beta = \gamma = \lambda = 1.$$

$$\frac{l_d}{d_b} = \frac{15 f_y \alpha \beta \gamma \lambda}{16 \sqrt{f_c'} \left(\frac{c + K_{tr}}{d_b}\right)} = \frac{15 \times 400 \times 1 \times 1 \times 1 \times 1}{16 \sqrt{28} \times 2.5} = 28.35$$

$$l_d = 28.35 \times 28 = 794 \text{ mm.}$$

Available Anchorage Length $= 825 - 50 = 775$ mm. Use hooks.

1

Reinforced Concrete Design (2)
Exam -2- / Solutions

① The cross-section shown in the figure is loaded by $V_u = 222$ kN and a torque $T_u = 40$ kN.m. Take $f_c' = 21$ MPa and $f_y = 400$ MPa.

(a) Check if torsion can be neglected.

(b) What is the required spacing of $\phi 12$ mm closed stirrups.

(c) What area of longitudinal steel is required for torsion.

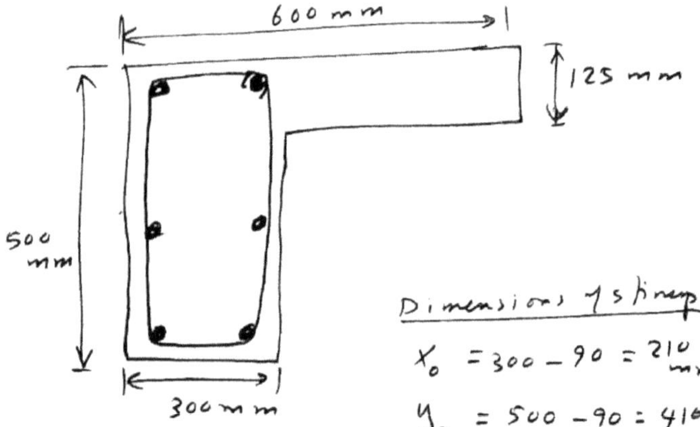

Solution:

$d = 500 - 40 - 10 - 13 = 437$ mm.

$b_w = 300$ mm.

Dimensions of stirrup:

$x_o = 300 - 90 = 210$ mm.

$y_o = 500 - 90 = 410$ mm

$P_{cp} = 600 + 125 + 300 + 375 + 300 + 500$

$= 2200$ mm.

$A_{cp} = (500 \times 300) + (300 \times 125) = 187,500 \text{ mm}^2$.

$P_h = 2(210 + 410) = 1240$ mm.

$A_{oh} = 210 \times 410 = 86,100 \text{ mm}^2$. (208 408)

(a) Check if torsion can be neglected:

$$\frac{\phi \sqrt{f_c'} A_{cp}^2}{12 P_{cp}} = \frac{0.85 \sqrt{21}\,(187,500)^2 \times 10^{-6}}{12\,(2200)} = 5.19 \text{ kN.m}.$$

but $T_u = 40$ kN.m $\not<$ 5.19 kN.m.

$\Rightarrow$ torsion <u>must</u> be considered.

2

(b): $V_c = \frac{1}{6}\sqrt{f'_c}\, b_w d = \frac{1}{6}\sqrt{21}\,(300)\left(\frac{437}{1000}\right) = 100 \text{ kN}.$

$\phi V_s = V_u - \phi V_c = 222 - 0.85\,(100) = 137 \text{ kN}.$

$V_s = \frac{\phi V_s}{\phi} = \frac{137}{0.85} = 161 \text{ kN}.$

$\frac{A_v}{s} = \frac{V_s}{f_y d} = \frac{161}{(400\times10^3)(437/1000)} = 0.000921053 \text{ m}^2/\text{m}$

$= 0.921053 \text{ mm}^2/\text{mm}.$

In terms of $\frac{A_t}{s}$, $\frac{A_t}{s} = \frac{1}{2}\frac{A_v}{s} = \frac{0.921053}{2} = 0.46053 \text{ mm}^2/\text{mm}.$

Compute A_t/s required for torsion with $\theta = 45^\circ$:

$A_o = 0.85\, A_{oh} = 0.85\,(86{,}100) = 73{,}185 \text{ mm}^2.$

$\frac{A_t}{s} = \frac{T_u}{2\phi f_{yv} A_o \cos\theta} = \frac{40}{2(0.85)(400\times10^3)(73{,}185\times10^{-6})\cos 45^\circ}$ ($\cot 45^\circ$)

$\Rightarrow \frac{A_t}{s} = 0.001137 \text{ m}^2/\text{m}$

$= 1.137 \text{ mm}^2/\text{mm}.$

$\frac{A_{t,total}}{s} = 0.46053 + 1.137 = 1.5972 \text{ mm}^2/\text{mm}.$

but $A_t = \frac{\pi(12)^2}{4} = 113 \text{ mm}^2.$

$\frac{113}{s} = 1.5972 \Rightarrow s = 70 \text{ mm}.$

maximum spacing is not to exceed 300mm or $P_h/8$:

$\frac{P_h}{8} = \frac{1240}{8} = 155 \text{ mm}.$

$\Rightarrow$ Use $\phi 12$ mm @ 70 mm c/c.
closed stirrups.

[3]

Check $\left(\frac{A_v}{s}\right)_{min} = \frac{0.35\, b_w}{f_y}$

$$\frac{A_v}{s} + 2\frac{A_t}{s} = 0.921053 + 2(1.137) = 3.195 \ mm^2/mm.$$

$$\left(\frac{A_v}{s}\right)_{min} = 0.35\frac{b_w}{f_y} = 0.35\frac{(300)}{400} = 0.2625 \ mm^2/mm$$

$\Rightarrow$ o.k.

(c) Compute longitudinal steel for torsion:

$$A_l = P_h\left(\frac{A_t}{s}\right) = (1240)(1.137) = 1410 \ mm^2.$$

Check the minimum area of longitudinal steel:

$\frac{A_t}{s}$ must not be less than $\frac{0.175\, b_w}{f_y} = 0.175\frac{(300)}{400} = 0.13125 \ mm^2/mm$

$$A_{l,min} = \frac{5\sqrt{f_c'}\, A_{cp}}{12 f_{yl}} - \left(\frac{A_t}{s}\right)p_h \frac{f_{yv}}{f_{yl}}$$

$$= \frac{5\sqrt{21}\,(187{,}500)}{12\,(400)} - (1.137)(1240)\left(\frac{400}{400}\right)$$

$$= -515 \ mm^2.$$

$\Rightarrow A_l = 1410 \ mm^2 > A_{l(min)} = -515 \ mm^2. \Rightarrow$ o.k.

---

(2). Design a square column footing for a 400-mm square tied interior column that supports a dead load $D = 890$ kN and a live load $L = 712$ kN. The column is reinforced with 8-Ø25 mm bars, the base of the footing is $D_f = 1.5$ m below grade, the soil weight is $\gamma_s = 1600\ kg/m^3$, $f_y = 400$ MPa, $f'_c = 21$ MPa, and the allowable soil pressure on the base of the footing is $q_a = 240\ kN/m^2$. L4

Solution:

(1) $D+L = 890 + 712 = 1602$ kN.

Assume a footing 0.6 m (600 mm) deep:

Net allowable soil pressure $q_n = 240 - (0.6 \times 25) - \frac{0.9 \times 1600 \times 9.81}{1000}$

$= 211\ kN/m^2$.

Required area $= \frac{D+L}{q_n} = \frac{1602}{211} = 7.59\ m^2$.

For a square footing, side $= \sqrt{7.59} = 2.75$ m.

⟹ Use 2.8 m.

Area $= 2.8 \times 2.8 = 7.84\ m^2 > 7.59\ m^2$.

$P_u = 1.4D + 1.7L = 1.4(890) + 1.7(712) = 2456$ kN.

(2) Design pressure $P_u = \frac{P_u}{Area} = \frac{2456}{7.84} = 313\ kN/m^2$.

$d = 600 - \underset{cover}{(75)} - \underset{d_b}{(25)} = 500$ mm.

5

③ Check punching shear:

$V_u = P_u - \text{soil pressure}$

$= 2456 - (313)(0.9)(0.9)$

$= 2202\ \text{kN}.$

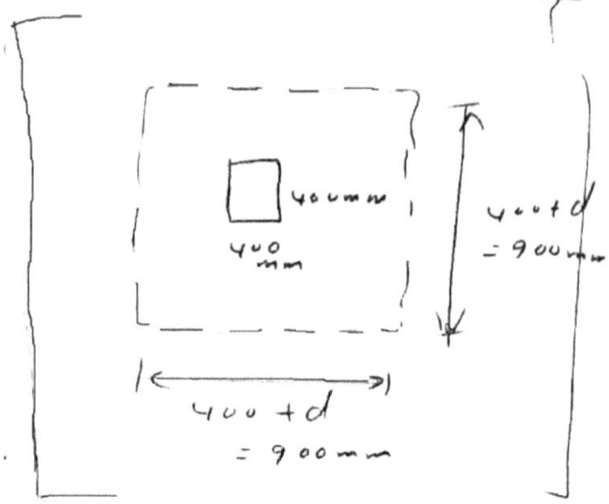

$b_o = 4(900) = 3600\ \text{mm}.$

Shear capacity equals the smaller of.

$\phi V_c = \frac{1}{3}\phi\sqrt{f_c'}\, b_o d = \frac{1}{3}(0.85)\sqrt{21}\,(3600)\left(\frac{500}{1000}\right) = 2337\ \text{kN}.$

or $\phi V_c = \phi\left(\frac{\alpha_s d}{b_o} + 2\right)\frac{1}{12}\sqrt{f_c'}\, b_o d = 0.85\left(\frac{40(500)}{3600} + 2\right)\frac{1}{12}\sqrt{21}\left(\frac{3600}{1000}\right)(500)$

$= 4416\ \text{kN}.$

$\Rightarrow$ Take $\phi V_c = 2337\ \text{kN}.$

$\Rightarrow V_u = 2202\ \text{kN} < \phi V_c = 2337\ \text{kN}.$

---

④ Check beam shear:

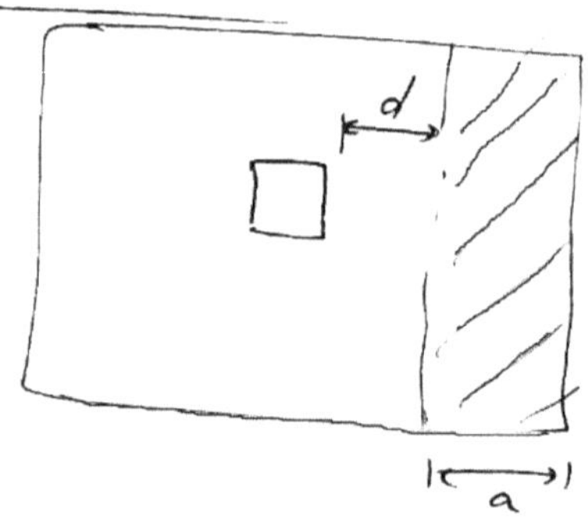

$a = \frac{(2.8\times1000) - 400 - 2(500)}{2}$

$= 700\ \text{mm}.$

$V_u = P_u\,(\text{area})$

$= (313)\left(\frac{700}{1000}\right)(2.8) = 613\ \text{kN}.$

$\phi V_c = \frac{1}{6}\phi\sqrt{f_c'}\, b_w d = \frac{1}{6}(0.85)\sqrt{21}\,(2.8)(500) = 909\ \text{kN}.$

$\Rightarrow V_u < \phi V_c \Rightarrow V_u = 613\ \text{kN} < \phi V_c = 909\ \text{kN}.$

<u>Note</u>: Footing depth can be reduced.

16

(5) Design the dowels (for the transfer of load from column to footing):

Bearing capacity of concrete $P_c = 0.85\,\phi f_c'\,A$

$$\Rightarrow P_c = (0.85)(0.7)(21)(400)\left(\frac{400}{1000}\right) = 1999 \text{ kN}.$$

Dowels $P_d = P_u - P_c = 2456 - 1999 = 457$ kN.

$$P_d = f_y A_s$$

$$457 = (400 \times 1000) A_s \Rightarrow A_s = 0.0011425 \text{ m}^2 = 1142 \text{ mm}^2.$$

$A_{s(min)}$ for dowels $= 0.5\%$ of column area

$$= \frac{0.5}{100} \times 400 \times 400 = 800 \text{ mm}^2.$$

$$\Rightarrow \text{Use } A_{s(dowels)} = 1142 \text{ mm}^2.$$

Use four dowels, $A_{s(\text{per dowel})} = \frac{1142}{4} = 285 \text{ mm}^2$.

Use $\phi 20$ mm bars, $A_{s(\text{supplied})} = \frac{\pi (20)^2}{4} = 314 \text{ mm}^2$.

Length of dowels:

$$l_{db} = \frac{f_y d_b}{4\sqrt{f_c'}} = \frac{(400)(20)}{4\sqrt{21}} = 436 \text{ mm}$$

but not less than $0.044 f_y d_b = 0.044(400)(20) = 352$ mm or 200 mm.

$$\Rightarrow \text{Use } l_{db} = 436 \text{ mm}.$$

$$l_d = l_{db}\left(\frac{A_{s,req.}}{A_{s,sup.}}\right) = 436\left(\frac{285}{314}\right) = 396 \text{ mm}.$$

Use $l_d = 400$ mm.

www.ingramcontent.com/pod-product-compliance
Lightning Source LLC
LaVergne TN
LVHW070858210926
848482LV00016B/966